Electrochemistry of Dihydroxybenzene Compounds

Electrochemistry of Dihydroxybenzene Compounds

Carbon Based Electrodes and Their Uses in Synthesis and Sensors

Hanieh Ghadimi
Department of Chemical and Biomolecular Engineering,
The University of Akron, Akron, OH, United States
Universiti Sains Malaysia, P. Pinang, Malaysia

Sulaiman Ab Ghani
Universiti Sains Malaysia, P. Pinang, Malaysia

IS Amiri
University of Malaya (UM), Kuala Lumpure, Malaysia

elsevier.com

Elsevier
Radarweg 29, PO Box 211, 1000 AE Amsterdam, Netherlands
The Boulevard, Langford Lane, Kidlington, Oxford OX5 1GB, United Kingdom
50 Hampshire Street, 5th Floor, Cambridge, MA 02139, United States

British Library Cataloguing-in-Publication Data
A catalogue record for this book is available from the British Library

Library of Congress Cataloging-in-Publication Data
A catalog record for this book is available from the Library of Congress

ISBN: 978-0-12-813222-7

Publisher: John Fedor
Acquisition Editor: John Fedor
Editorial Project Manager: Tasha Frank
Production Project Manager: Maria Bernard
Designer: MPS

Typeset by MPS Limited, Chennai, India

CONTENTS

LIST OF FIGURES

LIST OF TABLES

LIST OF SCHEMES

ACKNOWLEDGMENTS

Thanks to Professor Dr. Sulaiman Ab Ghani for his valuable guidance, motivation, patience, and constant supports throughout the completion of this book. We wish to extend deep thanks to Associate Professor Dr. Abdussalam Salhin Mohamed Ali and Professor Dr. Norita Mohamed for their supports.

LIST OF ABBREVIATIONS AND SYMBOLS

A_1	anodic peak
ACAP	acetaminophen
AcO	acetone
AA	ascorbic acid
ASA	acetylsalicylic acid
°C	degree centigrade
C_1	cathodic peak
CT	catechol
CHN	elemental analyses of total carbon, hydrogen, and nitrogen
CNTs	carbon nanotubes
CV	cyclic voltammetry
D	dimensional
DCM	dichloromethane
DHB	dihydroxybenzenes
dl	limit of detection
DMF	dimethylformamide
DPV	differential pulse voltammetry
DWCNTS	double-walled carbon nanotubes
δ	chemical shift
ΔE_p	peak separation
E_{pa}	anodic peak potential
E_{pc}	cathodic peak potential
EFTEM	energy-filtering transmission electron microscope
EIS	electrochemical impedance
EtOH	ethanol
F-test	statistical hypothesis test
FT-IR	Fourier transform infrared
FESEM	field emission scanning electron microscope
g	gram
GCE	glassy carbon electrode
GR	graphene
GO	graphene oxide
h	hour
HOAc	acetic acid
HQ	hydroquinone

Hz	hertz
I_{pa}	anodic peak current
I_{pc}	cathodic peak current
kHz	kilohertz
MΩ cm	megaohm centimeter
μA	microampere
mA	milliampere
mg	milligram
mHz	Megahertz
mL	milliliter
mm	millimeter
mmol	millimole
mM	millimolar
μL	microliter
μM	micromolar
MS	mass spectrometry
mV	millivolts
MWCNT	multiwalled carbon nanotubes
nM	nanomolar
NMR	nuclear magnetic resonance
Ω	ohm
PCT	paracetamol
P4VP	poly(4-vinylpyridine)
PNC	polymer nanocomposites
R	linear regression coefficients
R_{ct}	charge transfer resistance
R_s	solution resistance
RSD	relative standard deviation
S_B	standard deviation
SWCNT	single walled carbon nanotubes
SWV	square-wave voltammetry
THF	tetrahydrofuran
TLC	thin layer chromatography
TMS	tetramethylsilane
TSC	thiosemicarbazide
UA	uric acid
V	volt
$\nu^{1/2}$	square root of the scan rate
v/*v*	volume per volume
Z_w	Warburg element

CHAPTER 1

Introduction

1.1 NANOTECHNOLOGY AND NANOSCIENCE

Nanoscience and nanotechnology has become a very dynamic and critical area of research, which is dramatically developing and spreading as a general-purpose technology in almost every field of technology domain as well as science and engineering disciplines due to its ability to create superfunctional properties of materials at nanoscale [1,2].

Nanodimension transcends the conservative boundaries between scientific and engineering disciplines and technology segments. Apparently, nanotech is the foundation for gaining widespread benefits, consisting of smarter electronics, improved health, advanced agriculture, and cleaner source of energy [3]. The tendency to understand how materials behave when sample sizes are close to atomic dimensions is one of the most important motivations in nanoscience.

Applying nanostructures for technology is also an opportunity in order to miniaturize and use new applications and devices as a result of unique properties of nanostructures. It should be noted that although many nanostructures such as large molecules and quantum dots are of great interest, the most active areas of study are related to nanotubes. Nanoscience and nanotechnology primarily deal with the synthesis, characterization, exploration, and exploitation of nanostructured materials which have received considerable attention. Another reason for the great popularity of this field is that the nanoscale phenomena are of great interest to chemists, physicists, biologists, electrical and mechanical engineers, and computer scientists.

1.1.1 Type and Properties of Nanostructures

Nanomaterials consist of various types of nanostructured materials including clusters, quantum dots, nanocrystals, nanowires, and nanotubes, while collections of nanostructures involve arrays, assemblies, and super lattices of the individual nanostructures [4,5]. The properties of materials with nanometer dimensions significantly differ from those

Electrochemistry of Dihydroxybenzene Compounds. DOI: http://dx.doi.org/10.1016/B978-0-12-813222-7.00001-2

of atoms and bulk materials. This is mostly as a result of the nanometer size of the materials which causes (1) a large fraction of surface atoms; (2) high surface energy; (3) spatial confinement; and (4) reduced imperfections, which do not exist in the corresponding bulk materials [6]. The small dimensions of nanomaterials which provide them with an extremely large surface area to volume ratio will result in a large fraction of atoms of the materials to become the surface or interfacial atoms. However, when the sizes of nanomaterials are comparable to Debye length, surface properties of nanomaterials will affect the entire material [7,8]. This consecutively may enhance or modify the properties of the bulk materials. For instance, metallic nanoparticles could be used as active catalysts and chemical sensors fabricated from nanoparticles and nanowires which also have higher sensitivities and selectivities.

1.1.2 Applications of Nanomaterials

Nanotechnology is being applied practically in all fields ranging from science, engineering, and health to medicine. Nanomaterials provide unique physical, chemical, and mechanical properties that could be used for a wide variety of applications in different areas. Nanomaterials are receiving considerable interest due to the potential of applications in a wide range of fields such as catalysis, sensors, semiconductors, fuel cells, drug delivery, and functional coatings. They can also be used in cosmetics, textiles, medical diagnosis and therapeutics, healthcare, water, and air pollution treatment [9].

1.2 CARBON-BASED MATERIALS

Carbon is an essential constituent element of every living organism. We not only benefit from numerous carbon-based products in our life, but also use it as a scientific tool. The physical and chemical properties of carbon are well documented, including the use of carbon in nanotechnology which is a great interest area of research and considerable funding is being allocated to the carbon nanotechnology research. Graphene (GR) (sp^2 bonded), graphite, diamond (sp^3 bonded), carbon nanotubes (CNTs) (sp^2 bonded), fullerenes, and nanodiamonds are the well-known forms of carbon (Fig. 1.1) [10]. Nowadays, the applications of CNTs, GR, and carbon nanospheres in electrochemical sensors are the most interesting aspects of electrochemistry, and they are going to be very important in the future of electrochemistry and related technologies in general.

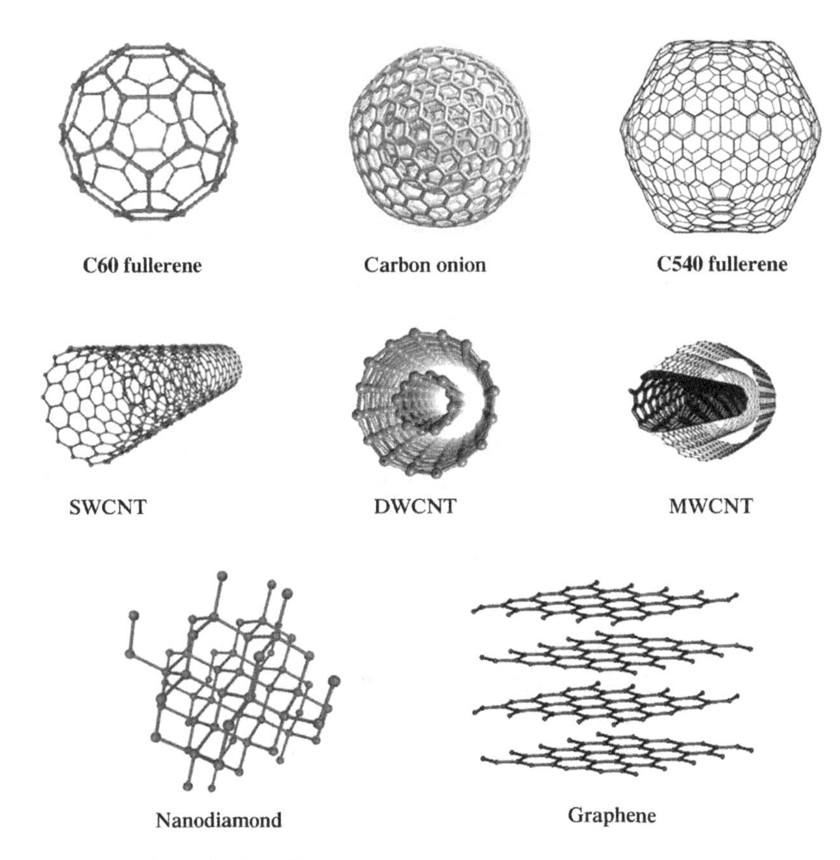

Figure 1.1 Various forms of carbon [12].

Carbon-based nanomaterials have received great attention regarding their electronic, mechanical, optical, and chemical characteristics. They can also be used for energy storage, molecular electronics, composite materials, bioengineering, and environment. The unique and tunable properties of carbon-based nanomaterials which are most commonly cited in environmental applications are size, shape, and surface area; molecular interactions and sorption properties; and electronic, optical, and thermal properties enable new technologies for identifying and addressing environmental challenges [11].

1.2.1 Carbon-Based Materials in Electrochemistry

Carbon-based materials such as glassy carbon, graphite, carbon fiber, graphite paste, and composite graphite have been broadly applied in a variety of electrochemical devices due to their excellent conductivity and chemical inertness. Good crystallinity and microstructural order has been shown by the natural graphite (electron configuration $2s^2$ $2p^2$) which consists of multilayers of GR. Therefore, it can be used as high-quality electrochemical devices with lower background current, wide potential window, fast electron-transfer rate and easily renewable surface. Thus, electrochemical devices fabricated from various-based materials have become more important, and their practical applications in electrochemistry have also been extensively studied with great success.

1.3 THE IMPORTANCE OF ELECTRODE SURFACE IN ELECTROCHEMISTRY

In voltammetry, the most essential parameters include the electron-transfer rate, background current, and accumulation of analytes. These parameters are significantly affected by the surface modification of the working electrode. Hence, the sensitivity and selectivity could be optimized by improving the surface phenomena of electrodes.

The carbon electrode is broadly used due to its numerous advantages in contrast to the metal electrode. First, it contains a majority of types which have been mentioned in Section 1.2.1. Second, the carbon electrode forms an abundance of surface oxides, which not only results in a lower background current but also in an acceleration of the transfer of the electron [10]. Generally, there are two surface types of carbon, including pristine and monolayer oxides [10]. The pristine surface which can be gained by heat or polishing is a disordered surface without oxides or impurities. The monolayer oxides which can be achieved by modifying a thick film of oxides are a rich oxygen-containing surface. Hence, the voltammetric background can be increased due to the increasing of double-layer capacitance of interface, and the adsorption of cationic species can be also enhanced by the modifier. As a result, the sensitivity and selectivity of electroanalysis are a great deal enhanced by this process. There is a large variety of modifiers used for electrochemical pretreatment of the electrode surface in order to modify the electrochemical performance between the surface and the bulk solutions. The coating of the electrode surface with nanoparticles is one of the most important methods used.

1.3.1 Multiwalled Carbon Nanotube

The CNTs have largely contributed in modern analytical sciences. As it is well documented, the basic composition of a CNT is GR, and the

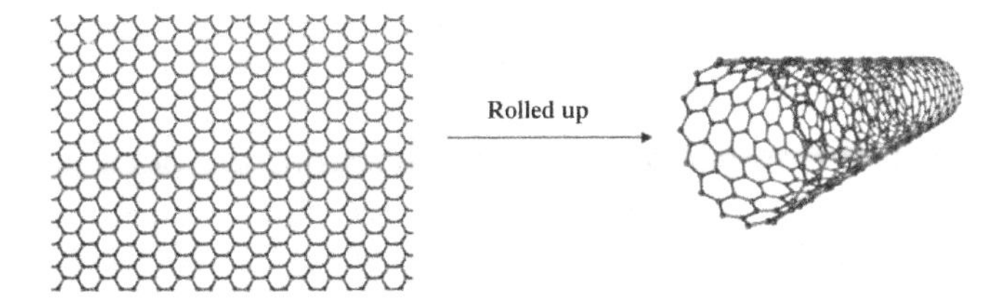

Figure 1.2 Graphene sheet rolled up to form CNT [13].

structure of a CNT is like a sheet of GR rolled up into a tube (Fig. 1.2). Besides their structural type, they are also classified as single-walled CNTs (SWCNTs), double-walled CNTs (DWCNTs), and multiwalled CNTs (MWCNTs) according to the layer numbers (Fig. 1.1). SWCNTs consist of only one layer of GR rolled to form a seamless cylinder with a few nanometers of diameter and a few hundred nanometers of length. DWCNTs consist of two GR cylinders with concentric arrangement and multi GR cylinders are termed as MWCNTs with diameters up to 100 nm [10].

The unique properties of CNTs have led to significant applications in several fields, such as in the electronics and medicine aerospace industries, which have also encouraged the need of analytical methodologies to characterize and control the quality of these nanomaterials. Moreover, the use of CNTs as analytical tools, and the construction of nanodevices and nanosensor based on CNTs are considered as other areas of development for modern analytical science.

The wide-ranging role of CNTs in analytical chemistry has been reported by Valcarcel et al. [14], Trojanowicz [15], and Merkoci [16]. Taking into consideration, in particular, the role of CNTs in electroanalytical chemistry, properties such as a high electronic conductivity and a high mechanical resistance have driven an impressive research effort in electroanalytical applications in recent years. Furthermore, an increased electrode active surface area, which gives rise to enhanced electrochemical responses, and a demonstrated antifouling capability of electrode surfaces upon modification with CNTs are other key practical advantages that have promoted numerous noteworthy applications in electroanalytical chemistry, including electrochemical sensors [17–19].

The ability to promote electron transfer reaction when used as an electrode in electrochemical reaction has been suggested by the special electronic properties of CNTs, which provides a new application in the electrode surface modification to design new electrochemical sensors and novel electrocatalytic materials [20].

CNT-modified electrode demonstrates the properties of electrocatalytic activity and electrosepration because CNT has unique electronic properties; thus, it can promote electron transfer reaction which could be applied in the detection of analytes in a low concentration or in the complex matrix. The CNT-modified electrodes generally consists of CNT-paste electrode, CNT-intercalated electrode, CNT-coating electrode, and CNT-embedded polymer electrode and are stated to give electrocatalytic activity toward a great variety of analytes, including dopamine [21], ascorbic acid (AA) [22], aspirin [23], acetaminophen (ACAP) [24], and various metal ions such as As^{3+}, Bi^{3+}, Pb^{2+}, and Cd^{2+} [25,26].

1.3.2 Graphene

GR as a novel carbon-based nanomaterial has aroused significant interest due to its potential applications in various fields [27–29]. GR is made from natural graphite which is found to be a new kind of nano-sized carbon-based material besides CNTs. It is the mother element of some carbon allotropes, including graphite, CNTs, and fullerenes (Fig. 1.1). GR is a flat and sheet-like form of graphite [30], and it is the basic form of CNTs. Structurally, GR is a one-atom-thick planar sheet of sp^2-bonded carbon atoms densely packed in a honeycomb crystal lattice. GR is not only considered as the basic form of CNTs, but also many other kinds of carbon-based materials. For instance, it has the ability to be rolled up to 1D CNTs, wrapped up to 0D fullerenes and stacked to 3D graphite [31,32]. Due to its sp^2 structure, GR has numerous advantages having unique electronic, mechanical, and optical properties. GR-based electrodes have demonstrated to have various applications such as transparent electrodes, ultrasensitive chemical sensors [33], supercapacitors (electric double layer capacitor), and nanoelectronics [34–36]. Among these mentioned properties, exploring its application in the field of electrochemical sensor is of particular interest for the current studies [37–49].

GR is made from chemical reduction of graphene oxide (GO), which may present the advantage of being cheap. More recently, electrochemical reduction of GO to GR has received great attention because it is fast, inexpensive, green, and nontoxic [50–52]. Fig. 1.3 illustrates the scheme of the oxidation/reduction process of GR.

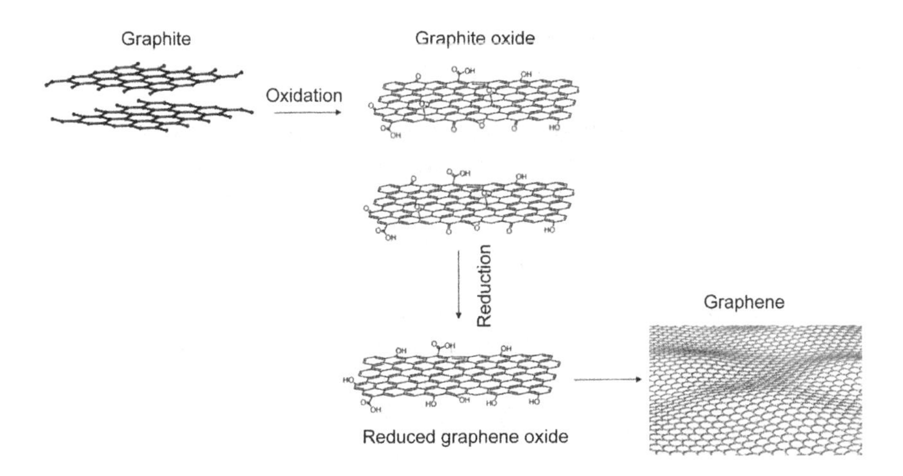

Figure 1.3 Production of GR via oxidation of graphite and reduction of graphite oxide to reduced graphene oxide (RGO) [53].

As can be seen from Fig. 1.3, the resulting reduced graphene oxide (RGO) from GO monolayers can be deposited in controllable density, and many conductive substrates can be applied. Thus, the preparation of thin RGO film from GO can be achieved onto various solid and flexible substrates with conductivity increasing four times of the substrates [54–56]. Therefore, the GR produced from reduced GO, seems friendly in its application in electrochemical sensors. GR-based electrodes have shown better advanced performance in terms of electrocatalytic activity compared to CNT-based ones. These findings pointed out that the opportunities in electrochemistry encountered by CNTs might be also available for GR [57]. Studies on GR are likely to provide a fundamental insight into all carbon materials. In comparison with CNTs, GR demonstrates potential advantages of low cost, high surface area, ease of processing, and safety [58].

Functionalization and dispersion of GR sheets are of critical importance in their applications. A large number of experiments focus on the insertion of additional chemical species between the basal planes of graphite [59,60]. GR, the building block of graphite, was theoretically established in 1940 [61]. GR sheets with a high specific surface are hydrophobic and have a tendency to form agglomerates which may limit its further applications [62]. Aggregation can be reduced by the attachment of other small molecules or polymers to the GR sheets. Therefore, the material which has the ability to prevent GR from aggregating has received growing interest nowadays. The presence of hydrophilic or hydrophobic groups prevents the aggregation of GR sheets by stronger polar–polar interactions or by their bulky size [63,64].

The functionalized GR sheets are more hydrophilic and can be easily dispersed in solvents with long-term stability [65,66]. In addition, proper chemical functionalization of GR for instance by conventional acid treatment method, for formation of –COOH and –OH groups, prevents the agglomeration of single layer GR. The aggregation can also be reduced by the attachment of other small molecules or polymers to the GR sheets [65,67,68]. Meanwhile, noncovalent functionalization, such as codispersion with polymers, has proved to be successful in solubilizing GR nanosheet [69]. The attachment of functional groups to GR also assists in its dispersion in a hydrophobic media as well as in the organic polymer. Therefore, an effective approach to the production of surface-functionalized GR sheets in large quantities has been a main focus of many researchers, with the goal of employing the majority of frequently proposed applications of GR in the areas of polymer nanocomposites (PNCs), solar cells, drug delivery systems, supercapacitor devices, memory devices, transistor devices, and biosensors [67]. As a novel and very capable material, it reveals fascinating applications in the fields of battery [70], supercapacitor [71], fuel cells [72], and ultrasensitive sensors [73]. In the area of electroanalytical chemistry, a great deal of attempts has been made to explore its electrocatalytic activity as electrode materials for the purpose of high sensitive analysis [39,74].

1.3.3 Benefits of Applying Carbon Nanotubes (CNTs) and Graphene in Electrochemical Analysis

It is very important to develop simple, sensitive, and accurate methods in order to identify pharmaceutical compounds and dihydroxybenzenes (DHBs) by the electrochemical method. It has been found that due to the physical and chemical properties of CNTs and GR, it tends easily to absorb organic molecules through van der Waals and π–π stacking interactions or noncovalent interactions. Therefore, once the CNT or GR film is coated on the surface of a bare electrode, those organic molecules are easily accumulated on the surface of the modified electrode. The concept of chemically modified electrodes is one of the exciting developments in the field of electroanalytical chemistry. Several different strategies have been employed in the modification of electrode surface. The motivations for the modifications of the electrode surface area include improved electrocatalysis, freedom from surface fouling, and prevention of undesirable reactions competing kinetically with the desired electrode process [75]. Some examples of the most ordinary methods are mentioned briefly in the following sections.

1.4 POLYMER NANOCOMPOSITE (PNC) BASED ON CARBON NANOMATERIAL ELECTRODE

PNCs are of great interest for a variety of applications due to its numerous features such as low weight, low cost, ease of processing and shaping, and corrosion resistant [76]. PNCs are polymers that have been reinforced with small quantities (less than 5% by weight) of nanosized particles having high aspect ratios [77]. The specific development of polymeric nanocomposite based on conventional polymers and conductive carbonaceous material has received a great deal of attention as a route to obtain the new materials with new structural and functional properties superior to those of the pure components and of previous nanocomposite systems with other fillers [78]. These properties could include increased modulus and strength, outstanding barrier properties, improved solvent and heat resistance, and decreased flammability [79].

1.4.1 PNC Based on Multiwalled Carbon Nanotube (MWCNT)

CNTs with exceptional structural properties and the presence of edge plane like sites located at the end and in the defect areas of their tubules possess high electrical conductivity, high chemical stability, and extremely high mechanical strength. Due to its excellent properties, CNTs can be used as ideal reinforcing agents for high performance polymer composites. They have been applied broadly in the fabrication of ion sensors and biosensors with the least fouling, reduced overvoltage effects, faster electron transfer kinetics and higher sensitivities than the traditional carbon electrodes [80,81]. However, poor solubility and process ability of CNTs have been the key technical obstacles, hindering their biomedical, and other promising applications. It is extremely hard to disperse MWCNT in liquid. There are several methods for the dispersion of nanotubes in the polymer matrix consisting of solution mixing, melt mixing, electrospinning, in situ polymerization, and chemical functionalization of the CNT. Also several types of CNT-modified electrodes have been studied viz. CNT paste electrode [82], CNT film-coated electrode [83,84], CNT power microelectrode, and conducting polymer/CNT-modified electrode [85,86]. The latest is of special interest because they have three-dimensional structures due to the incorporation of CNT into conducting polymers.

The noncovalent and covalent modifications of the CNTs with polymers are generally applied to improve their dispersion and orientation in aqueous solutions [87–89]. The noncovalent approach mostly includes surfactant modification [90,91], polymer wrapping [92], and polymer absorption [93,94]. The interaction between the aromatic moieties of the polymers and the CNT surface were proposed to be responsible for the introduction of polymers into the CNT samples [95]. Among the noncovalent methods, polymer wrapping is the simplest route. Polymer/CNT nanocomposite is one of the most promising field for CNT [96,97]. These PNCs have remarkable enhanced properties in electrical, thermal, physical, chemical, conductive, and other smart functionalities which have attracted great attention worldwide from both academic and industrial points of view.

1.4.2 PNC Based on Graphene

Material scientists are examining materials with improved physicochemical properties that are dimensionally more appropriate in the field of nanoscience and technology. In this respect, the discovery of GR and GR-based PNCs is an essential addition in the nanoscience area having a key role in modern science and technology [98]. The discovery of PNCs by the Toyota research group has created a new dimension in the field of material science.

GR is expected to have significant properties such as outstanding conductivity, high specific surface area; high mechanical, thermal, and chemical stabilities compared to CNTs and electrically conducting reinforced nanocomposites [99–103]. Mentioned properties provide GR as a promising material that could be employed in many applications such as in photovoltaic devices, transparent electrodes, sensors, super capacitors, and conducting composites [98,104–109]. The inherent properties of GR have created great interest in its possible implementation in innumerable devices [110] leading to future generations of high speed and radio frequency logic devices, thermally and electrically conducting reinforced nanocomposites, ultra-thin carbon films, sensors, electronic circuits, and transparent and flexible electrodes used for displays and solar cells [110].

It should be noted that the tensile strength of GR is similar or slightly higher than that CNT and the thermal and electrical conductivity of GR is also higher than those of CNT. The superior properties of GR compared to polymers are also reflected in polymer/GR nanocomposites. Polymer/GR nanocomposites present better mechanical, thermal, gas barrier and electrical properties compared to the neat polymer. Various polymer matrices have been employed for composites, including thermoplastics [111–113], liquid crystalline polymers [114,115], water soluble polymer [116], and conductive polymers [117] among others.

1.5 CONDUCTIVE POLYMER

Conductive polymers are a new category of materials which demonstrate highly reversible redox behavior. Electrically conducting polyacetylene was discovered by scientists in 1977 [118,119] and the Chemistry Nobel Prize was awarded to the scientists in 2000 [120]. The forthcoming utility of conductive polymers with an effective application in the growing technologies in bimolecular electronics display devices and electrochemical storage system. The most important conductive polymers is polyenes or polyaromatics such as polyaniline [121], polypyrrole [122], and polythiophene [123]. These compounds have many potential applications as well as their derivatives, including sensors, biosensors, electrochromic devices [124] electromagnetic shielding [125,126], corrosion inhibitor [127,128], super capacitors [129,130], polymeric batteries [131], and polymeric actuators [132,133]. Intrinsically, these conducting polymers are important owing to their relatively high thermal stability, low costs, ease of synthesis, and good environmental behavior among the polymer-modified electrodes. Conducting polymer-based drug delivery systems which is used as a therapeutic electrode material for microfabricated neural devices have developed much and dramatically during the past few decades [128].

1.5.1 Types of Conducting Polymers

Types of conducting polymers fall under three headings [134]:

1. Intrinsically/inherently conducting polymers.
2. Conducting polymer composites.
3. Ionically conducting polymers.

Polyacetylene, polypyrrole, polythiophene, polyaniline, poly(*p*-phenylene), and poly(phenylene vinylene) are the most common examples

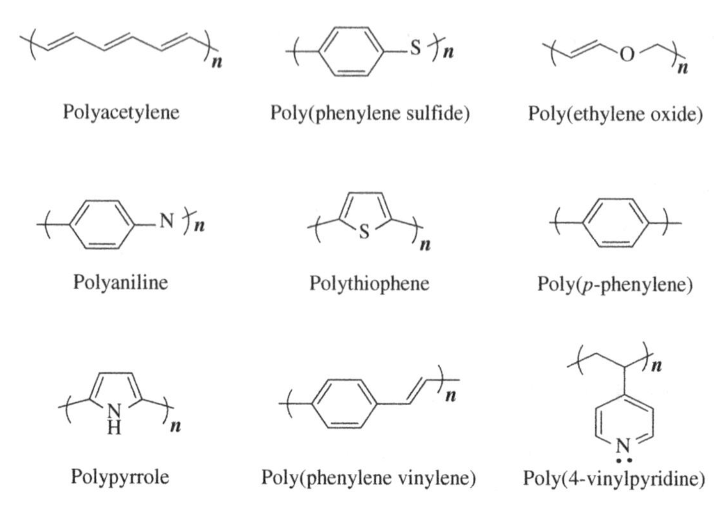

Figure 1.4 Examples of inherently conducting polymers.

of intrinsically/inherently conducting polymers. Fig. 1.4 illustrates some of the conjugated polymers studied as intrinsically conducting polymers. Simplicity of synthesis, low cost, good environmental stability, and high conductivity are some properties which makes conducting polymer desirable for use.

The two main groups of applications for organic conducting polymers are briefly described in the following section:

Group I: This group utilizes polymers conductivity as its main property including its light weight, biological compatibility for ease of manufacturing or cost, electrostatic materials, conducting adhesives, printed circuit boards, electromagnetic shielding, artificial nerves, antistatic clothing, active electronics (diodes and transistors), piezoceramics, and aircraft structures. Conducting polymer is also incorporated in the construction of super capacitors with efficient charge propagation [135]. Efficient super capacitors can be applied as power conditioners in cameras, in power generators and also as excellent power supplies.

Group II: This group uses the electroactivity character property of the materials consisting of molecular electronics, electrical displays, optical computers, ion exchange membranes, and electromechanical actuators. Moreover, one of the other important applications of these polymers is in the field of sensors such as chemical sensors, ion, and biochemical sensors [136] and in the medical field owing to the fact that several tissues in the human body arc responsive to electrical fields [120]. Most conducting polymers have high biocompatibility and the ability to trap and release biological molecules. Hence, this makes them useful as tissue-engineering scaffolds, neural probes, bioactuators, and also drug delivery devices [137]. Among the mentioned conducting polymers in Fig. 1.4, poly(4-vinylpyridine) (P4VP) is of interest in this research due to its electrical conductive properties and good redox polymer.

1.5.2 Poly(4-vinylpyridine) (P4VP) as a Conducting Polymer

The use of conducting polymers such as P4VP and polypyrrole to improve the electrocatalytic activity, electron transfer kinetics and stability of the modified electrodes have been reported [138–142]. P4VP is a hydrophobic polymer in a polar solvent and is a cationic (protonated) polyelectrolyte at low pH [142]. It has been reported that noncovalent and covalent modifications of the carbon-based nanomaterials, such as CNTs and GR, with polymers or biomolecules are aimed at improving their dispersions and orientations in aqueous solutions. The flexible chains of P4VP may act as a stabilizing agent when wrapped around the MWCNT or GR, preserving its intrinsic electrical and mechanical properties and also rendering their solubility in water, ethanol and dimethylformamide (DMF). Furthermore, P4VP can easily be processed and fabricated to form free-standing films which are desired in most applications [143]. This is an important feature for assembling CNT and other materials applying the layer by layer assembly approach [144]. Besides that, the high electrical conductivity and good redox mediator are among the analytical advantage of P4VP in PNC electrodes [138,145,146].

1.6 ELECTROORGANIC SYNTHESIS

Among the most interesting and innovative chemical technologies, the electrochemical method provides a very appealing and potent means for the small-scale and high purity production of compounds. From the point of view of green chemistry, using electrosynthesis is considered as a method having significant advantages. The electrochemistry of organic compounds was being studied for more than 150 years with the first one suggesting applicability of the development of green methods organic synthesis [147]. The benefit of applying the electroorganic synthesis method is well documented in previous research by

Nematollahi group [148–150] and others [151–153], which confirmed that the emergence of organosynthesis via electrochemical procedures, were especially advantages of the electrochemistry. These include energy specificity, clean synthesis, absence of toxic reagents and solvents, operation at room temperature and pressure, chemical selectivity and specific activation of small molecules [152].

Electrosynthesis can result in efficient and occasionally unexpected synthesis of compounds, which cannot be easily prepared by conventional organic synthesis [154]. The top green advantages of the electroorganic synthesis are listed as follows:

1. The electroorganic reaction is effectively achievable at room temperature.
2. Since the electrode potential has the ability of being controlled over a wide range, a wide variety of electroorganic reactions are able to be designed.
3. Since the structures and reactivities of the active species are different from those formed in usual chemical reactions, a diversity of novel reactions can be realized.
4. The electroorganic reaction is fundamentally nonpolluting.
5. The reaction scale is controlled easily.
6. The energy efficiency is usually high.

1.7 DIHYDROXYBENZENES AND ITS DERIVATIVES

Ortho- and *para-*DHBs are well known in biological systems. 1,2- and 1,4-naphthalenediols are often used as metabolites of simpler aromatic hydrocarbons such as benzene or naphthalene. Due to the existence of two exchangeable hydrogen atoms, the aromatic diols is likely to be biologically reactive molecules, capable of demonstrating both anti- and prooxidant behavior. These compounds have antioxidant and antimicrobial activity and they are usually important elements of plant extracts used in medicinal chemistry for their various biological activities [155]. Some examples of the DHBs compound are described briefly in the following section.

1.7.1 Catechol

Catechols (CTs) such as DHB are antioxidants (Fig. 1.5). They have mixed effects on normal and cancer cells at in vitro and in vivo.

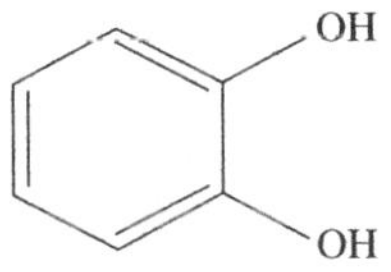

Figure 1.5 The structural formula of CT.

Depending on the amount of DHB used and the time before its application, DHB could reduce tumor growth [156]. Due to their mechanisms [157–162], synthesis [148–152,155,163–165] and the importance of interest, the study of reaction between quinones and nucleophiles is increasing. Selection of solvents, pH, nature of nucleophiles and electrophiles and type of DHB will affect the mechanism. CT is also an important material which is widespread in the environment, and especially as constituents of edible plants [166] it is also widely used in tanning, cosmetic, chemical materials, and in pharmaceutical industries. CT is hazardous to human health occurring frequently in the environment [167]. In addition, CT is one of the most important phenolic compounds which occur naturally in fruits and vegetables and which could be released in the environment with its manufacture and use. It is also detected at low levels in ground water, soil samples, and in wastewater from coal conversion [168]. CT and their quinone derivatives have received significant attention due to their abundance in nature and key roles in many biological systems. Furthermore, most of the compounds incorporating CT moiety exhibit antioxidant, anticarcinogenic, antifungus, and antibacterial activities are applied as HIV integrase inhibitors.

1.7.2 Hydroquinone

Hydroquinone (HQ) (1,4-DHB), which is one of the isomers of diphenol (Fig. 1.6), is very vital in a large number of biological and industrial processes such as paper manufacturing, coal-tar production, and photography [169]. HQ has been widely used in dye production for skin whitening, cosmetic products, antioxidant, polymer, pharmaceutical industries and as an anticancer agent. HQ is a potentially carcinogenic substance, causing severe effects on the central nervous system [170]. Indeed, HQ is a serious environmental pollutant for instance exposure to HQ can cause irritation to the eyes, nose, throat, skin, and nail discoloration [171]. Therefore, the determination and quantification of HQ is necessary to estimate its possible effect. Recently,

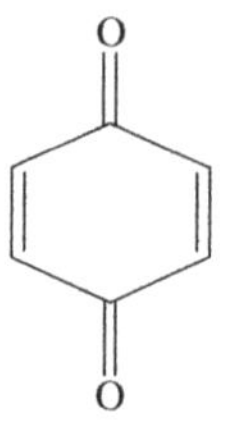

Figure 1.6 The structural formula of HQ.

utilizing HQ for substitution of unstable oxygen is such growing interests in order to act as electron transfer media in construction of enzyme-based glucose biosensor [172].

1.8 ELECTROCHEMICAL SYNTHESIS OF CT IN THE PRESENCE OF NUCLEOPHILE

CT as one of the most well-known compounds in biological systems frequently performs as the reactive center of electron transfer in the structure of numerous natural compounds. An enormous number of quinones with various structures are provided by nature and have a very important role in the redox electron-transport chain of living systems [173]. It is well documented that CT could be oxidized electrochemically to *o*-quinone and since the quinones formed are quite reactive they could be attacked by a variety of nucleophiles, such as methanol [174], 4-hydroxycoumarin, 4-hydroxy-6-methyl-2-pyrone [175], sulfinic acids [159,176,177], barbituric acids [178], and were converted to the corresponding methoxyquinone [174], coumestan [175], spiro [179], dispiropyrimidine benzofuran [178], and arylsulfonylbenzendiol [176,177] derivatives, respectively. Moreover, since electrochemical oxidation is often similar to the oxidation of CT in the mammalian and human body's central nervous system, this appealing incident leads to the study of anodic oxidation of CT in the presence of different nucleophile [174]. Some examples of the most important electroorganic synthesis methods which have been investigated in the previous research [166,180] are discussed in the following discussion.

The electrooxidation of 4-methylcatechol in the presence of 1,3-dimethylbarbituric acid and 1,3-dimethylthiobarbituric acid as nucleophiles has been studied in depth by CV and controlled-potential coulometry. The results reveal that 4-methylcatechol would be oxidized to *o*-benzoquinone with no conversions to its quinone methide tautomeric form, under Electrochemical chemical electrochemical chemical (ECEC) [181,182] mechanism. The electrochemical synthesis of the final product has been successfully performed in one pot in an undivided cell [166].

Fotouhi et al. [183] have worked on the electrochemical oxidation of CT in the presence of 6-methyl-1,2,4-triazine-3-thion-5-one in aqueous sodium acetate, using CV and controlled-potential coulometry. They suggested a mechanism for the oxidation of CT and their reaction with 6-methyl-1,2,4-triazine-3-thion-5-one. The results show that CT derivatives were converted into 7*H*-thiazolo [3,2-*b*]-1,2,4-triazin-7-one derivatives under a Michael-type addition reaction of 6-methyl-1,2,4-triazine-3-thion-5-one to anodically generated *o*-quinones. The results also confirmed the successful performance of electrochemical synthesis of some new 6-methyl-1,2,4-triazine-3-thion-5-one derivatives in an undivided cell.

Electrochemical oxidation of CT in the presence of cyanoacetone and methyl cyanoacetone as nucleophiles in aqueous solutions has been investigated using CV and controlled-potential coulometry under ECEC mechanism. New nitrile containing benzofurans has been found as the final products [184].

The electrochemical oxidation of CT in the presence of sulfanilic acid was investigated using CV and controlled-potential coulometry [159]. The results illustrate that the *o*-quinone derived from CT take part in a Michael addition reaction with the sulfanilic acid.

Electrochemical oxidation of quercetin which is a flavonoid broadly spread in nature has been studied in the presence of benzensulfinic acids as nucleophiles in an acetonitrile water mixture, using CV and controlled-potential coulometry. The results revealed that the *o*-quinone derived from quercetin was involved in a Michael addition reaction with the benzenesulfinic acids in order to form the corresponding sulfonyl derivatives, consequently the electrochemical synthesis of these derivatives has been effectively performed at carbon rod electrode in an undivided cell in good yield (74%) and purity [185].

Electrochemical oxidation of 4-aminocatechol has been examined in the presence of 4-toluensulfinic acid and benzensulfinic acid as nucleophiles in aqueous solutions, via CV and controlled-potential coulometry. The results show that the *o*-quinone derived from 4-aminocatechol takes part in Michael addition with these nucleophiles in order to form the corresponding new organosulfone derivatives [180]. The electrochemical behavior of CT and 4-methylcatechol in the presence of methyl mercapto thiadiazol as nucleophiles has also been examined [152] using electroorganic synthesis through controlled-potential coulometry.

From all the above discussion, there has been no study done about the electrochemical oxidation of CT in the presence of thiosemicarbazide (TSC). The importance of the mentioned motivating compounds has encouraged us to investigate the electrodecarboxylation of CT in the presence of TSC.

1.8.1 Thiosemicarbazide

Organosulfones are considered as important intermediates in organic synthesis due to their chemical properties [186] and biological activities [187]. Diarylsulfones are also important synthetic targets which are widely used synthons for synthetic organic chemists in their diverse applications which are very useful in the practice of medicinal chemistry as the sulfone functional group is used in numerous drugs. Furthermore, some of sulfone compounds have been shown to inhibit HIV-1 reverse transcriptase and represent an emerging class of substance having the ability to address toxicity and resistance problems or nucleoside inhibitors [188]. One of these classes is TSC (Fig. 1.7). Owing to their structural features, they have potent biological activity [189], particularly as antimicrobial and antifungal agents [190]. The biological activities of TSC could be enhanced by the functional groups on the parent aldehyde or ketone [191]. The concentration on TSC derivatives is justified since in vitro biological evaluation indicates that these derivatives have antitrypanosomal, antiamoebic activity in noncytotoxic concentrations to mammalian cells [192]. TSC is also good as a potential antitumor [193], antimalarial, antiviral, antibacterial, radioprotector, anticonvulsant agent, ulcer inhibitor, and anticancer agent [193].

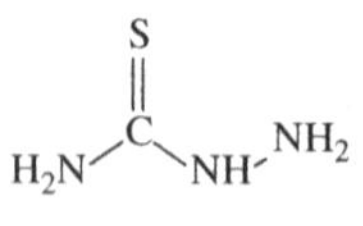

S
C
H_2N NH NH_2

Figure 1.7 The structural formula of TSC.

1.9 METHODS USED FOR DETERMINATION OF DIHYDROXYBENZENE (DHB)

Several methods have been developed to determine DHB (CT and HQ). These include chromatography [194], spectrophotometry [195], capillary electrophoresis [196], chemiluminescence [197], pH-based flow injection analysis [198], and electrochemical methods [199–201]. The electrochemical methods have their advantages in various aspects such as excellent sensitivity, high accuracy, good reliability, and low cost of instrumentation compared to the other modern techniques. However, since both HQ and CT are electrochemically active, they can be easily determined using electrochemical techniques.

1.9.1 Electrochemical Method

The determination of phenolic compounds such as DHB is one of the great interests to many fields such as industrial analysis, biological, and environmental control [202,203] due to their excellent electrochemical activities.

1.10 ELECTROCHEMICAL SENSORS FOR ANALYSIS

Sensor science and engineering is relevant to almost any aspects of life including safety, security, surveillance, monitoring, and awareness in general. Sensors are also playing a key role in industrial applications in order to process control, monitor, and ensure safety. Sensors also play a critical role in medicine that is being applied in diagnostics, monitoring, critical care, and public health. Electrochemical sensors have been employed broadly either as a whole or an integral part of a chemical and biomedical sensing element. In addition, the use electrochemical sensors to measure analytes are of great interest in clinical chemistry and are ideally suited for these new applications due to their high sensitivity and selectivity, portable field-based size, rapid response time, and economical aspect. The different principles are required in a particular design of the electrochemical cell. Electrochemical sensors generally could be classified as conductivity/capacitance, potentopmetric, amperometric, and voltammetric sensors [204]. The amperometric and voltammetric sensors are characterized by their current-potential relationship with the electrochemical system which so far, there has been little definition about these sensors. Electrochemical sensors could be fabricated to enormously small dimensions as well and consequently they become ideal for direct placement.

Nanomaterials have become very popular in the latest electrochemical sensing research as a result of their electrical conductivity, unique structural and catalytic properties, high loading of biocatalysts, good stability, and excellent penetrability [205]. Carbon is one of the most extensively-used material in electroanalysis and electrocatalysis. Carbon-based electrodes have been generally used on account of some aspects which consist of economical cost, good electron transfer kinetics and biocompatibility.

Recently, CNTs have also been incorporated into electrochemical sensors [84,206] which offer distinctive advantages consisting of enhanced electronic properties, a large edge plane/basal plane ratio, and fast electrode kinetics. Generally, CNT-based sensors have superior sensitivities, lower detection limits, and faster electron transfer kinetics than traditional carbon electrodes.

Testing and optimizing several variables is required in order to create a CNT-based sensor. The performance is dependent on the synthesis method of the CNT, CNT surface modification, the method of electrode attachment, and the addition of electron mediators. CNTs have revealed significant performance in biosensors [207–209], biofuel cells [210], and polymer electrolyte membrane fuel cells [211–214]. GR-based electrodes have presented greater performance which consists of electrocatalytic activity [215], good biocompatibility, and conductivity than CNTs based ones.

GR-based nanomaterials have lately revealed interesting applications in electrochemical sensors and biosensors, which offer a valuable sensing platform for small biomolecules [38,216]. Furthermore, the electrochemical properties of GR could be effectively modified by combining with other functional nanomaterials such as catalyst nanoparticles and composite material to create versatile electrochemical sensing performance [45,217–220].

1.10.1 CT as a Sensor for Electrochemical Determination

CT derivatives have been used as electron transfer mediators in electrochemical processes due to their high electron transfer efficiency, excellent redox reversibility, and low cost [221]. These compounds as mediators are immobilized on the electrode surface by various methods, such as adsorption mixing into carbon paste electrode, or simply added to the test solution. On the other hand, the oxidation of CT provides reactive species which sulphydryl thiols will readily bind through 1,4-addition reaction [222]. White et al. [223] have noted the electrochemical oxidation of CT at the glassy carbon electrode (GCE) in the presence of cysteine. Their results showed that the deposition of a quinine–cysteine adduct was adsorbed on the electrode surface. The electrochemically initiated reaction of aromatic amines with CT which has been reported by Seymour et al. [224] has revealed a facile direction to the electrochemical determination of the former.

1.10.2 Electrochemical Sensor for Detection of CT and HQ

A variety of voltammetric techniques based on modified electrodes materials such as carbon nanostructures and polymers have been offered to enhance detection selectivity to CT in the presence of HQ. Nanomaterials are of great interest to the field of sensors and catalysis owing to their unique electronic, catalytic, and optical properties.

However, to the best of our knowledge, there is no report on the use of P4VP/MWCNT-modified GCE (P4VP/MWCNT–GCE) and P4VP/GR nanosheet composite modified GCE (P4VP/GR–GCE) as electrode modifier in the determination of CT and HQ.

1.10.2.1 Determination of Catechol

According to the mentioned CT properties in Section 1.7.1, developing the selective analysis methods for CT is a crucial issue. Chen et al. [167] studied selective determination of CT in wastewater at silver doped polyglycine-modified film electrode. The results revealed that a novel method for the determination of CT through CV was confirmed and the modified electrode revealed a significant electrocatalytical effect on the redox of CT. Xu et al. [225] have investigated electrocatalytic oxidation of CT at MWCNT-modified electrode. The electrochemical behavior of CT by cyclic voltammetric techniques at the screen printed graphite electrode and its application for the determination of CT in a water sample has been reported by Mersal [168].

1.10.2.2 Determination of Hydroquinone

Numerous analytical techniques have been developed consisting of high performance liquid choromatography [195,226], capillary electrochromatogrphy, flow injection analysis [227], spectrophotometry [228], and electrochemistry [169]. Li et al. [229] have reported sensitive and selective determination of HQ on the surface of GR-modified/GCE (GR/GCE) which was applied to detect HQ in water samples with a low detection limit of 0.8 μM. Yiyi et al. [230] stated the electrochemical determination of HQ using hydrophobic ionic liquid-type carbon paste electrodes. Kavanoz and Pekmez [171] have examined the amperometric determination of HQ by poly(vinylferrocenium) perchlorate-polyaniline composite film-coated electrode. A novel, simple, sensitive, and reliable method for determination of HQ based on β-cyclodextrin/poly(*N*-acetylaniline)/CNT composite β-CD/PAA/MWNTs film modified GCE has been reported by Kong et al. [169]. Sensitive detection of HQ was viable and stability, and reproducibility of the modified electrode was also reported.

1.10.2.3 Simultaneous Determination of CT and HQ on Modified Electrodes

DHB are nowadays the major pollutants in environmental waters. They are difficult to degrade and this makes them very harmful even at very low concentrations [231]. The development of quick, simple, and precise methods for the analysis of DHB compounds is then imminent. However, simultaneous determination of HQ and CT are not very successful. Simultaneous detection of HQ and CT is highly desirable as they usually coexist in products due to their similarities in structures and properties [203,229]. The main problem arises from overlapping of redox peaks of the corresponding isomers, and in many cases, it is a significant barrier for most conventional solid electrodes. A chemically modified electrode is an excellent approach to address this problem [232]. The modifier must be capable of improving the peak separation between these isomers. These PNCs have remarkable enhanced properties in electrical, thermal, physical, chemical, conductive, and other smart functionalities which have attracted great attention worldwide from both academic and industrial points of view. For example, the oxidation of organic compounds and other phenolic compounds such as HQ and CT is greatly enhanced once the electrode is modified by coupling CNTs with some conducting polymers [233,234].

In the following discussion, some examples of the most relevant literature on the simultaneous determination of CT and HQ are mentioned briefly.

A sensor based on the CNT-ionic liquid composite for simultancous determination of CT and HQ has been developed by Bu et al. [235]. Their results show that the selectivity of electrochemical method for CT and HQ has been improved at the modified electrode. Moreover, no significant interference from common cationic species is observed. This proposed method is also applied in food detection, medical, and environmental control with multicomponent analysis.

In another study by Ahammad et al. [233], the simultaneous determination of CT and HQ applying poly(thionine)-modified GCE (PTH-modified GCE) has been also investigated. This modification process was found to be simple compared to other electrode modification methods stated elsewhere for sensitivity and selectivity. The suggested mechanism and the signal separation of CT and HQ at the PHT-modified GCE were also discussed. Good reversibility and catalytic activity for the electrochemical redox reaction for these two diphenols were accomplished at the PTH-modified GCE.

Wang et al. [236] studied the simultaneous electrochemical determination of CT and HQ in binary mixtures at a poly(phenylalanine)-modified GCE as well. The elimination of the fouling effect by the oxidized product of these two mentioned diphenols isomers on the response of CT has been reported as one of the other achievements at the poly(phenylalanine)-modified electrode. The developed electrode is suitable for simultaneous determination of CT and HQ in the water sample with high selectivity.

Zhang et al. [234] developed a direct simultaneous determination method for CT and HQ by CV and differential pulse voltammetry (DPV) at the MWCNT/poly(3-methylthiophene)/GCE. The higher peak current on the MWCNT/poly(3-methylthiophene)/GCE was also obtained which clearly demonstrated the MWCNT/poly(3-methylthiophene)/GCE as an efficient promoter to improve the kinetics of the electrochemical process of CT and HQ. The proposed electrode in their study proved to have a better application prospect for the highly selective quantitative determination of CT and HQ separately as well as in the mixture commonly found in water samples.

Direct simultaneous electrochemical determination of CT and HQ at a poly(glutamic acid)-modified GCE was examined by Wang et al. [203]. Fast electron transfer, high selectivity, and outstanding sensitivity for the oxidation of CT and HQ were attained at the poly(glutamic acid)-modified electrode.

Simultancous determination of CT and HQ at single wall CNT and electrochemically activated GCE has been performed by Wang et al. [237] and Kong et al. [238]. Results revealed that the modified electrode demonstrate a considerable response for DHB isomer electrooxidation.

The voltammetric behaviors of DHB isomers were examined at an ordered mesoporous carbon modified GCE (OMC/GC) by Bai et al. [239]. The OMC-modified electrode showed a significant electrocatalytic activity toward DHB isomer.

1.11 APPLICATION OF THE NANOCOMPOSITE-MODIFIED ELECTRODES FOR PHARMACEUTICAL ANALYSIS

Drug analysis plays a key role in the quality control of drug formulations having a great impact on public health. Pharmaceutical substances are responsible for the relief of pain and symptoms caused by diseases. However, they could be hazardous or even fatal when adulterated or used excessively. Therefore, the development of appropriate analytical methods for drug-quality control is considered crucial in successfully treating diseases, minimizing side effects, and complying with government oversight. A simple, sensitive, and accurate method to determine the active ingredients in drugs seems essential [240]. Consequently, numerous electrochemical techniques and modified electrode methods have been discussed in the previous studies. Certain drug combinations such as paracetamol (PCT), aspirin, and caffeine are used for treatment which mainly tend to maximize the pharmacodynamic interaction of compounds [241].

1.11.1 Determination of Paracetamol (PCT)

PCT or ACAP (Fig. 1.8) as a painkiller is broadly used for headaches, backache, and postoperative pain [242]. It was first synthesized in the late 19th century and is considered to have analgesic and antipyretic properties [243]. Taking high doses of PCT may result in undesirable effects in the body, although in appropriate doses, it does not present any side effects and with regard to its wide uses due to its remarkable therapeutic characteristics, accurate determination, and control of its quality is essential [244].

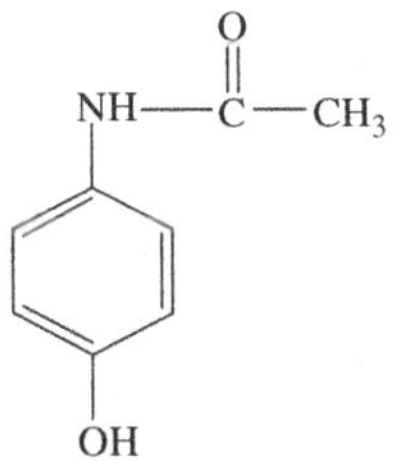

Figure 1.8 The structural formula of PCT.

A variety of methods have been applied to determine PCT in pharmaceutical formulations and biological fluids which consists of electroanalytical [244,245], capillary electrophoresis, spectrophotometry [246], infrared spectroscopy [247] liquid chromatography [248], and HPLC methods [249,250]. The majorities of the mentioned methods need pretreatment of samples and are also tedious and troublesome, while electroanalytical methods are more preferred due to the several factors described earlier. The selectivities of these methods are usually high as the analyte can be easily traced by its standard potential.

The features and benefits of modified GCEs in the manufacturing of ion sensors and biosensors and their electroanalytical applications have been reported [242,251]. Regarding the properties mentioned earlier, Habibi et al. [252] used single-walled CNT-modified carbon ceramic electrode (SWCNT/CCE) for the simultaneous determination of ACAP and AA. The results also revealed that selective determination of ACAP and AA is not interfered by some biological species and common inorganic ions. The developed electrode was confirmed to be useful for the determination of ACAP and AA in real pharmaceutical and biological samples with satisfactory results.

Beitollahi et al. [24] have reported simultaneous determination of isoproterenol, ACAP and *N*-acetylcysteine on the surface of carbon paste electrode modified with 5-amino-3′,4′-dimethyl-biphenyl-2-ol/CNT. The advantages of the examined modified electrode consist of high sensitivity and low detection limit, together with the ease of preparation and surface regeneration and high repeatability and stability of the modified electrode.

Electrochemical oxidation and sensitive determination of ACAP in pharmaceuticals at screen printed electrodes modified with electrogenerated poly(3,4-ethylene-dioxythiophene) film (SPE/PEDOT) has been examined by Su and Cheng [253]. The results demonstrated PEDOT efficiently accelerating the ACAP electron transfer rate and the redox behavior of ACAP at SPE/PEDOT is a diffusion-controlled quasi reversible process. The suggested electroanalytical methods were successfully used for quantitative determination of commercial pharmaceuticals containing ACAP alone or as a mixture with caffeine.

Kang et al. [66] have studied sensitive detection of PCT at GR-based electrochemical sensor. They also investigated the electrochemical behaviors of PCT on GR-modified GCEs by applying CV and square-wave voltammetry (SWV) revealing that the GR-modified electrode presented excellent electrocatalytic activity to PCT.

1.11.2 Determination of Acetylsalicylic Acid (ASA)

Acetylsalicylic acid or ASA (Fig. 1.9), another pharmaceutical product having antiinflammatory, antipyretic, and analgesic properties, is broadly used in the treatment of fever, headache stemming from cold, renal function [254] and is also effective in Alzheimers disease [255,256], cancer [257], and cardiovascular illness [258,259].

As was mentioned in Section 1.11.1 for PCT, different analytical methods have also been used for the determination of ASA. Recently, the determination of ASA using electrochemical methods was found to receive much attention due to its sensitivity and simplicity.

Lu and Tsai [23] examined electrocatalytic oxidation of ASA at the MWCNT-alumina-coated silica (MWCNT–ACS) nanocomposite modified GCE by applying CV and SWV. The MWCNT–ACS nanocomposite modified GCE was also successfully used in the determination of ASA in pharmaceutical samples with SWV. The results of CV and SWV revealed higher current responses and lower oxidation potential of ASA which indicated the electrochemical oxidation of ASA was improved by the electrocatalytic activity of (MWCNT–ACS) nanocomposite.

HO O
O O
H

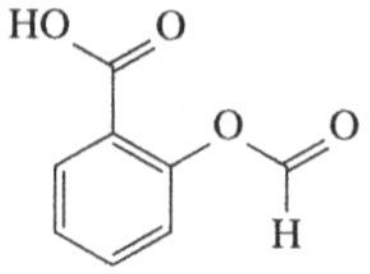

Figure 1.9 The structural formula of ASA.

The electrochemical behavior of ASA has been investigated at a platinum electrode in aqueous solutions by Wudarska et al. [260]. The process of oxidation and its kinetics has also been examined by CV and DPV.

1.11.3 Determination of Caffeine

Caffeine or trimethylxanthine (Fig. 1.10) is an alkaloid from the xanthine group which performs as a stimulant to the cardiac muscle central nervous system, resulting in sleeplessness [261]. Caffeine is applied recreationally as well as medically in order to reduce physical exhaustion and restore alertness when drowsiness occurs. It also results in sleeplessness, increased focus, faster and clearer flow of thought, and better general body coordination. It also comprises several physiological effects such as gastric acid secretion and dieresis [262]. Sometimes, it is also included in analgesic preparations due to its diuretic action.

Different methods have been used to determine caffeine including spectrophotometric [263] and chromatographic methods [264]. The mentioned methods proved to have many disadvantages such as high cost, time consuming, and they are more complicated than the electroanalytical methods.

Mersal [265] has investigated the electrochemical behavior of caffeine by applying simple, cheap and highly selective pseudo carbon paste electrode. In this study, SWV was applied for direct electrochemical determination of caffeine and the effect of different experimental parameters has been studied on the peak height of caffeine. A lower detection limit of 3.48×10^{-7} M was obtained. The suggested method confirmed to be useful for the direct electrochemical determination of caffeine in different real samples.

Voltammetric behavior of caffeine in beverages with significant sensitivity and selectivity applying a Nafion-ruthenium oxide pyrochlore chemically modified electrode has been reported by Zen et al. [261].

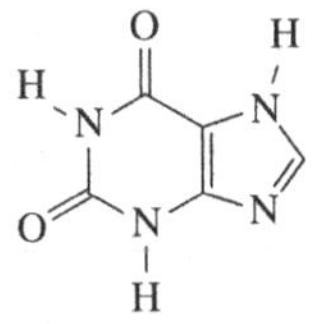

Figure 1.10 The structural formula of caffeine.

Compared to a bare GCE, the chemically modified electrode demonstrates a remarkable enhancement of the current response. The detection limit in the mentioned study was 2 μM.

In 2012, Faria et al. [266] have studied simultaneous determination of ASA and caffeine in pharmaceutical formulation applying a boron-doped diamond electrode by DPV. Owing to the high electrode stability and excellent repeatability, good resolution was attained between the drug oxidation peaks.

Sanghavi and Srivastava [267] revealed that a CNT paste electrode modified in situ with Triton X-100 that has been developed for the individual and simultaneous determination of ACAP, ASA, and caffeine. These studies demonstrated the oxidation of ACAP, ASA, and caffeine is simplified at an in situ surfactant-modified MWCNT paste electrode.

1.12 PROBLEM STATEMENT

In an organic reaction, the activation of a substrate molecule has generally been performed by the donation of photoenergy or thermal energy from outside the reaction system [147]. On the other hand, in an electroorganic reaction the formation of an active species from a substrate is achieved through transfer of the electrons between the substrate and an electrode [147].

The current study aims at using electroorganic synthesis method based on its benefits and significances including:

- Achievable at room temperature;
- Ability to be controlled over a wide range of electrode potential resulting in designing a wide variety of electroorganic reactions;
- A variety of novel reactions may be realized as the structures and reactivities of the active species are different from those formed in bulk chemical reactions;
- The possibility of enhancing the electroorganic reaction by using mediators;
- Being nonpolluting, having usually high energy efficiency the reaction scale can be easily controlled.

Voltammetric peaks corresponding to oxidation/reduction of two phenol isomers such as CT and HQ are, in many cases, highly overlapped; thus, a significant barrier for most conventional solid electrodes. A chemically modified electrode is an excellent approach to address the signal separation problem.

These DHBs are among the major pollutants in environmental waters nowadays. They are difficult to degrade which make them very harmful even at very low concentrations. The synergistic effects of MWCNT and P4VP in the P4VP/MWCNT nanocomposite and GR and P4VP in the P4VP/GR nanocomposite have provided an effective microenvironment for the redox process of these diphenols.

Since PCT, AA, and uric acid (UA) have nearly the same redox potential range, it is difficult to detect these compounds on the surface of unmodified electrode. The developed electrodes can be applied for detection of PCT in the presence of AA and UA.

Electroanalytical methods have rarely been used for the analysis of ASA and caffeine, mainly because their oxidation occurs at a very positive potential. As a result, a simple differential pulse voltammetric method using a composite-modified electrode could be used for quantitative determination of caffeine and ASA.

1.13 OBJECTIVES

This study aims to synthesize new DHB derivatives by electrochemical methods and to contribute to the related current electrochemical studies on the compounds. The study focuses on developing a simple, highly sensitive, and accurate voltammetric method to determine diphenols and other important pharmaceutically compounds using composite-modified glassy carbon-based electrodes. This research intends to achieve the following objectives to realize the mentioned aims:

- To investigate electrochemical oxidation of CT in the presence of nucleophile;
- To synthesize and characterize new DHB derivatives by electrochemical methods;
- To investigate on the combination of (1) MWCNT and P4VP and (2) GR nanosheet and P4VP as modifiers in the fabrication of modified GCE;
- To get the optimum parameters for the best performance electrodes;

- To develop a simple and highly sensitive electrochemical method for the detection of DHB and other pharmaceutically compounds with a P4VP/MWCNT-modified GCE (P4VP/MWCNT−GCE) and P4VP/GR nanosheet composite-modified GCE (P4VP/GR−GCE) based on the anodic peak currents of the corresponding analytes;
- To determine on the quality and efficiency of the modified electrode in the analysis.

CHAPTER 2

Experimental

2.1 MATERIALS

Catechol (CT), hydroquinone (HQ), thiosemicarbazide (TSC), ethylthiosemicarbazide, paracetamol (PCT), aspirin (ASA), caffeine, urea, thiourea, and 2,4-dinitrophenylhydrazine were purchased from Sigma-Aldrich Chemical Co, USA. Acetic acid (HOAc), sulfuric acid, hydrochloric acid (36%), acetonitrile (MeCN), acetone (AcO), ethanol (EtOH, 99.5%), methanol (MeOH), tetrahydrofuran (THF), chloroform ($CHCl_3$), dimethylformamide (DMF), and dichloromethane (DCM) were supplied by Merck, Germany. Aniline was obtained from Sigma-Aldrich, USA. Sodium acetate ($CH_3COONa \cdot 3H_2O$), potassium dihydrogen phosphate (KH_2PO_4), dipotassium phosphate (K_2HPO_4), aluminum powder, and other salts were from Sigma-Aldrich, USA. The poly(4-vinylpyridine) (P4VP) cross linked with 6% ethylene dimethacrylate (mesh size 50 μm) was supplied by Fluka Chemie, Switzerland. All chemicals were used without any further purification. The multiwalled carbon nanotubes (MWCNT) were purchased from Chengdu Organic Chemicals Co. Ltd., Chinese Academy of Sciences, China. Graphene nanopowder was purchased from Graphene Supermarket, Calverton, NY. All solutions were freshly prepared with pure water (18.2 MΩ cm).

2.2 INSTRUMENTS

All electrochemical experiments were performed using a potentiostat/galvanostat EG & G model 273 A (Princeton Applied Research, USA) and electrochemical workstation BAS Epsilon (Bio analytical system, USA). A conventional three-electrode system, including a platinum wire as an auxiliary electrode and Ag/AgCl (3 M NaCl) as a reference electrode, were used (Fig. 2.1). The working electrode was either unmodified glassy carbon disc (GCE) (3 mm in diameter) or GCE modified with P4VP/MWCNT or P4VP/graphene (GR). In controlled-potential coulometry, an assembly of five composite graphite

Electrochemistry of Dihydroxybenzene Compounds. DOI: http://dx.doi.org/10.1016/B978-0-12-813222-7.00002-4

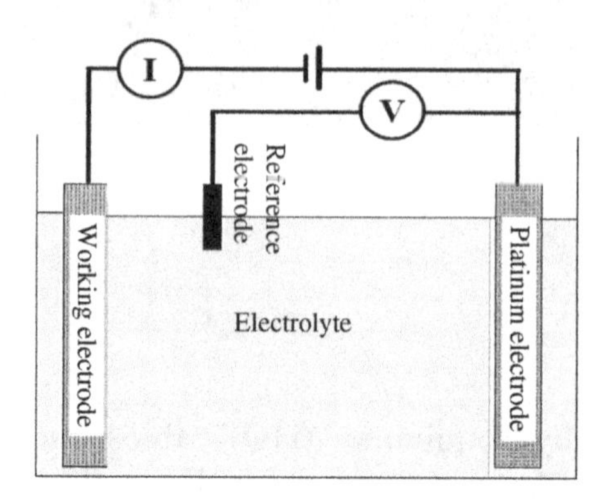

Figure 2.1 An electrochemical cell used for the electrochemical measurement.

rods (6 mm in diameter and 4 cm in length) and platinum plate (2 cm × 2 cm) was used as working electrodes and as the counter electrode, respectively. Unless otherwise stated, all potentials were obtained against the reference electrode. Impedance measurements were performed using EG & G FRD 100 (Princeton Applied Research, USA). The results of electrochemical impedance spectroscopy (EIS) were analyzed by using Zsimp Win 3.22 software. The pH was measured using a pH combination electrode model 915600B (Orion Res Inc, USA).

The synthesized product was characterized by IR, ^{1}H NMR, ^{13}C NMR, and elemental analysis. The NMR spectrum was obtained by Bruker Avance III; ascend TM (500 MHz). IR spectra were recorded using a Perkin-Elmer 2000 FT-IR spectrometer with an average of 64 scans in the frequency range of 400–4000 cm^{-1} with a resolution of 4 cm^{-1}, using KBr system. The analysis for carbon, hydrogen, and nitrogen was carried out using CHN/O Analyzer (Perkin-Elmer Series II 2400, USA). The surface and quantitative morphologies of the P4VP/MWCNT–GCE and P4VP/GR–GCE were studied using a field emission scanning electron microscope (FESEM) model Leo Supra 50 VP (Carl Zeiss, Germany) and the energy-filtering transmission electron microscope LIBRA 120 equipped with an Olympus SIS ITEM Version 5.0 (build 1243) (Carl Zeiss, Germany). All measurements were conducted at 25 ± 5°C. The pure water was collected from Milli-*Q* plus (millipore, USA). All sonication were performed in a Branson Ultrasonic Cleaner (USA). Centrifuge was carried out using KUBOTA 5920 (Japan).

2.3 ELECTROCHEMICAL METHOD FOR SYNTHESIS OF DIHYDROXYBENZENES DERIVATIVES

2.3.1 Preparation of Working Electrode

The composite graphite electrode was prepared in four steps. First, the composite graphite was washed in methanol for 30 min by sonication and then washed with distilled water, air dried and polished with sandpaper. Prior to use, the GC electrode was polished with alumina then rinsed thoroughly with water, cleaned in an ultrasonic bath with water, and finally rinsed with deionized water.

2.3.2 Electrochemical Study of CT in the Presence of TSC

The experiments were carried out by studying the cyclic voltammetric behavior of the reagent in 0.1 M acetate buffer (CH_3COOH/CH_3COONa) (pH 3, 4, and 5.3) and phosphate buffer (KH_2PO_4/K_2HPO_4) (pH 6, 6.5, 6.7, and 7.1) solutions as supporting electrolytes and acetonitrile as the solvent at potential range of −0.25 to 0.8 V versus Ag/AgCl (3 M NaCl) electrode. The cyclic voltammogram (CV) was scanned at different rates between 20 and 500 mV s^{-1}. Acetonitrile (2 mL) was added as a cosolvent due to the poor solubility of TSC in water. The CV of 1 mM CT and 1 mM TSC in an acetate buffer (pH 5.3) was recorded in a potential range between −0.25 and 0.8 V versus Ag/AgCl (3 M NaCl) electrode. For the determination of TSC, different concentrations of TSC (1×10^{-5} to 8.5×10^{-4} M) were added to the 1×10^{-3} M CT, and the CV was then recorded after each addition. The electrolyte solution was purged with oxygen free nitrogen gas to eliminate oxygen before use.

2.3.3 Electroorganic Synthesis

In 80 mL 0.1 M acetate buffer (pH 5.3), a mixture of 2 mmol CT and 2 mmol TSC in a mixture of H_2O–acetonitrile (9:1, v/v) was electrolyzed on composite graphite rods (6 mm o.d., and 4 cm in length) and a large platinum plate cathode at 0.38 V versus Ag/AgCl (3 M NaCl) electrode. The electrolysis process was interrupted several times and the graphite anode was washed in acetone for reactivation. The electrolysis was stopped when the decay of the current become more than 95%. The end of electrolysis was obtained through the consumption of starting material, checked by thin layer chromatography using ethyl acetate and petroleum ether (v/v: 30:70). At the end of the electrolysis, a few drops of hydrochloric acid were added to the solution and the cell was then placed in a refrigerator overnight. The formed solid was filtered. Then the solution was extracted with DCM. The separated organic layer was dried using $MgSO_4$, filtered and evaporated using rotary evaporator. The resulting solid product was characterized by FT-IR, ^{1}H NMR ^{13}C NMR, CHN, and MS analysis.

2.4 ELECTROCHEMICAL STUDY OF MODIFIED ELECTRODES

2.4.1 Purification and Acetic Functionalization of MWCNTs

The MWCNT was purified using the following procedure: first, it was purified by sonication in 6 M HCl for 4 h. Then, 50 mg of purified MWCNT was sonicated in a 40 mL mixture of H_2SO_4/HNO_3 (v/v = 3:1) in a water bath at 40°C for 6 h. After cooling to ambient temperature, the MWCNT was centrifuged, collected, and washed with pure water to neutral (pH 7). The final solid was then dried in oven (55°C) to a constant weight.

2.4.2 Preparation of P4VP/MWCNT-Modified GCE

A liquid phase blending method [268] was used to make a thin film composite. The modified electrode was fabricated as follows:

The GCE was polished with alumina slurry on the petri dish then rinsed thoroughly with a mixture of water/ethanol (v/v = 2:1), cleaned in an ultrasonic bath and finally rinsed with deionized water. The clean GCE was then dried in ambient temperature. The mixture of P4VP and MWCNT in a weight ratio of 2:4 was dispersed in 1 mL DMF for 3 h in an ultrasonic bath.

The P4VP/MWCNT–GCE was prepared by a drop casted method [268]. A 10 μL of P4VP/MWCNT dispersion (1.0 mg mL^{-1}) was dropped onto the surface of GCE and dried at 25 ± 5°C. The resulting P4VP/MWCNT–GCE was then used for further experiments. Fig. 2.2 shows the schematic diagram of the modified electrode.

2.4.3 Graphene Sheets Functionalization

The GR sheets were functionalized by mixing with a mixture of nitric acid and sulfuric acid (1:3 v/v) prior to sonicating in a water bath sonicator for 2 h at 40°C. The mixture was then washed using deionized water and centrifuged at 1000 rpm for 10 to 30 min to remove the residual acids in the supernatant. The washing step was repeated

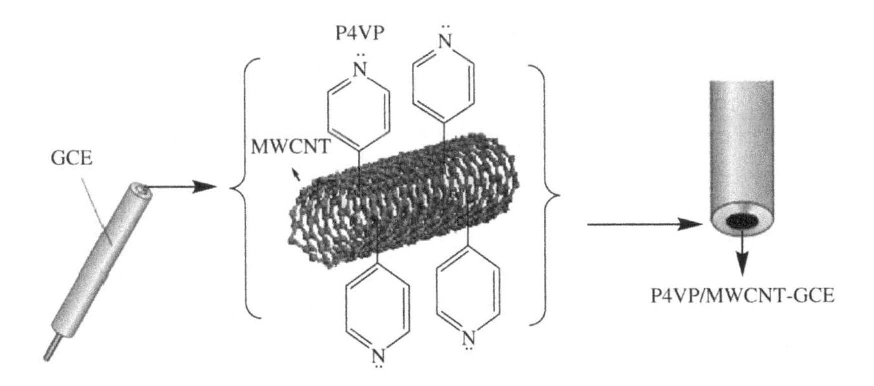

Figure 2.2 Schematic-modified electrode.

until the pH of the mixture was neutral. The final solid was then dried in an oven (55°C) to a constant weight.

2.4.4 Preparation of P4VP/GR-Modified GCE

The P4VP/GR-modified GCE was fabricated using the same procedure described earlier (Section 2.4.2). The GCE was polished with alumina and then cleaned in an ultrasonic bath, first in ethanol and then in water for a period of 15 min each. The clean GCE was dried at ambient temperature. The mixture of GR and P4VP in a weight ratio of 4:2 was dispersed in 1 mL DMF for 3 h. A 15 μL of P4VP/GR solution (1.0 mg mL^{-1}) was drop-casted onto the surface of GCE and dried at room temperature. The resulting nanocomposite P4VP/GR–GCE was then used for further experiments.

2.5 DETERMINATION OF TSC IN REAL SAMPLES

Real water samples for analysis were collected from Harapan Lake, USM. The samples were first filtered to remove suspended particles before use. The pH of the filtrate is 7.3. Then, the samples were kept in the refrigerator.

2.5.1 Determination of TSC in Water Samples

A 0.0045 g (1 mM) of TSC was dissolved in 2 mL acetonitrile and then diluted to the 50 mL with acetate buffer (pH 5.3). A 0.0027 g CT was dissolved in 25 mL same buffer solution (pH 5.3). The synthetic water samples were prepared by adding known amounts (2×10^{-4} M) of TSC to 25 mL real water samples (tap and lake water, separately). The CV on the response of TSC in the water samples for the CT reduction wave with subsequent additions of TSC was then studied. The concentration of TSC was quantified by the anodic peak currents of CT dependent on the concentration of TSC [224].

2.5.2 Determination of TSC in Propranolol Tablets

A stock solution of propranolol was made by dissolving 0.0088 g of propranolol tablets (Rouz Darou, Tehran—Iran) after grinding to a fine powder in 25 mL of distilled water. A (1 mM) stock solution of TSC was prepared by dissolving 0.0045 gm TSC in 2 mL acetonitrile and diluted with acetate buffer (pH 5.3) to 50 mL. The stock solution of CT was prepared by dissolving 0.0027 g of CT was in 25 mL acetate buffer (pH 5.3).

The electrochemical measurements for TSC in propranolol tablets were conducted by adding known amount (2×10^{-4} M) of TSC to 25 mL propranolol solution. The solution was then diluted to 50 mL with 0.1 M acetate buffer (pH 5.3). A portion of the resulting solution (25 mL) was then used as the sample for the determination of TSC. CV on the response of TSC in the propranolol sample for the CT reduction wave with subsequent additions of TSC was studied. The concentration of TSC was measured by the anodic peak currents obtained.

2.6 ELECTROCHEMICAL SENSOR STUDIES

2.6.1 Electrochemistry Determination of Diphenols

The electrochemical determination of diphenols using P4VP/MWCNT–GCE and P4VP/GR–GCE was carried out by studying the cyclic voltammetric behavior of the reagent in 0.1 M sodium sulfate buffer solution (pH 2.5) as supporting electrolyte at a potential range of −0.25 to 1 V. The CV was scanned at different rates between 10 and 400 mV s^{-1}. The differential pulse voltammetry (DPV) was performed with potentials from −0.25 V to 0.8 V with a step potential of 2 mV, modulation amplitude of 50 mV and scan rate of 10 mV s^{-1}. Experiments were conducted at 25 ± 5°C. All potential measurements were carried out against the Ag/AgCl (3 M NaCl) reference electrode.

2.6.2 Electrochemistry of PCT

The electrochemical determination of PCT using P4VP/MWCNT−GCE and P4VP/GR−GCE were investigated by studying the cyclic voltammetric behavior of the reagent in phosphate buffer (pH 7) as supporting electrolyte and buffer at a potential range of 0.0 to 0.8 V. The CV was swept at scan rates between 10 and 400 mV s^{-1}. The DPV was performed with potentials from 0.0 to 0.8 V, a step potential of 2 mV, modulation amplitude of 50 mV and a scan rate 10 mV s^{-1}. Experiments were conducted at 25 ± 5°C. All potentials were measured against reference electrode Ag/AgCl (3 M NaCl).

2.6.3 Electrochemistry of Aspirin and Caffeine

A 0.0450 g (5 mM) of ASA was dissolved in 2 mL acetonitrile and placed into a 50 mL standard volumetric flask and then diluted to the mark using phosphate buffer (NaH_2PO_4, Na_2HPO_4) (pH 7.4). The solution was then transferred into the micro-electrochemical cell where the measurements were carried out. Cyclic voltammetric experiments of ASA using P4VP/MWCNT−GCE were investigated by scanning the potential from 0.25 to 2 V. The CV was scanned at different rates between 10 and 300 mV s^{-1}. The DPV was performed with potentials from 0.6 to 2.7 V, a step potential of 2 mV, modulation amplitude of 50 mV and a scan rate 10 mV s^{-1}. Experiments were conducted at 25 ± 5°C. All potentials were measured against reference electrode Ag/AgCl (3 M NaCl). The electrochemical determination of 3 mM caffeine was conducted using the same procedure applied for ASA.

2.7 REAL SAMPLE ANALYSIS

2.7.1 Determination of Diphenols in Water Samples

A 0.0011 g (0.1 mM) of diphenols was diluted to 100 mL with 0.1 M sodium sulfate buffer (pH 2.5). The water samples were prepared by adding known amounts (2 μM) of HQ and CT to tap and lake water samples. The determination of HQ and CT in the samples was carried out using DPV at the modified electrode between −0.25 and 0.8 V. The concentrations of HQ and CT were quantified by the anodic peak currents.

2.7.2 Determination of PCT in Formulation Tablets

The developed electrode was tested for the determination of PCT in tablets (Pharmaniaga, Malaysia). The tablets were ground into powder and mixed with a mortar. A portion of the sample (equivalent to 0.05 g PCT) was accurately weighed and dissolved in 25 mL of distilled water. The solution was then diluted to 50 mL with 0.1 M phosphate buffer (pH 7). A portion of the resulting solution (25 mL) was then used as the sample for the determination of PCT using DPV.

2.7.3 Determination of PCT in Urine Mid-Samples

The use of P4VP/MWCNT−GCE is investigated for the measurement of PCT in three human urine mid-samples. Recovery tests were carried out by adding 25 μM of PCT standard solution to the diluted urine samples of healthy specimens of human. The urine samples were collected 4 h after the introduction of a 0.5 g PCT tablet. The untreated samples (250 mL) were then collected and diluted 30 times with phosphate buffer (pH 7).

2.7.4 Determination of ASA in Formulation Tablets

The developed electrode was tested for the determination of ASA in ASA tablets (Y.S.P Industries (M) Sdn. Bhd, Malaysia). The tablets were weighed, ground into powder, and then dissolved in 25 mL of distilled water. The solution was transferred to a 50 mL volumetric flask and diluted with 0.1 M phosphate buffer (pH 7.4). A portion of the resulting solution (25 mL) was then used as the sample for the determination of ASA concentration using DPV. The standard addition method was used for the sensitive determination of ASA in tablets.

2.8 CHARACTERIZATION OF THE MODIFIED ELECTRODES

2.8.1 Electrochemical Characterization

The cyclic voltammetry of 5 mM $[Fe(CN)_6]^{3-/4-}$ in 0.5 M KCl of the modified electrodes were carried out over a potential range of −0.25 to 0.8 V. The EIS measurement was performed after cyclic voltammetry subsequently, and the results were plotted as Nyquist plots with a frequency range from 100 kHz to 100 mHz and AC voltage amplitude of 5 mV.

2.8.2 Morphology Characterization

2.8.2.1 Field Emission Scanning Electron Microscopy (FESEM) Study

To obtain a secondary electron image of organic and inorganic materials with nanoscale resolution, the topographical and morphological studies were carried out using FESEM. In this study, the FESEM with an acceleration voltage of 15 kV was used for the characterization of the modified electrodes. Samples were carbon-coated to eliminate charging effects.

2.8.2.2 Transmission Electron Microscopy

In this book, the surface morphology of the P4VP/MWCNT and P4VP/GR were characterized by TEM, using a 120 kV Carl Zeiss Libra 120, Germany. Samples in DMF were ultrasonicated for 20 min to obtain a homogeneous suspension. A drop of suspension in DMF was placed on a carbon-coated copper grid, followed by solvent evaporation. The images were taken at random locations of sample to ensure the images recorded are representative as a whole. The images were taken with several magnifications in order to obtain information about the sample in general and also a closer visualization.

CHAPTER 3

Results and Discussion

3.1 CYCLIC VOLTAMMETRIC STUDIES OF CT IN ABSENCE AND PRESENCE OF TSC

To investigate the cyclic voltammogram (CV) behavior of catechol (CT), the measurements were conducted in the absence and presence of thiosemicarbazide (TSC). The CV of 1 mM CT at steady state are shown (Fig. 3.1). Fig. 3.1a shows that CT has an anodic peak potential (E_{pa}) at 0.439 V and a cathodic peak potential (E_{pc}) at 0.174 V. According to previous literatures [150,269], the transformation of CT (**1a**) to *o*-benzoquinone (**2a**) and vice versa (Scheme 3.1) is a two-electron and two-proton process. The peak separation (ΔE_p) is higher than 200 mV versus Ag/AgCl (3 M NaCl) which indicates that the electrode process is quasireversible [161,270]. As the peak−current ratio (I_{pa}/I_{pc}) is nearly unity, the electrode process is diffusion-limiting. The quinine layer formed on the surface of the electrode is stable. In the time scale of the experiment, any hydroxylation or dimerization reactions are not expected to occur, as they are too slow [271,272]. When one equivalent amount of 1 mM TSC was added, the height of the anodic peak A_1 for the oxidation of CT is increased, whereas that of the catholic peak C_1 for the reduction of CT is decreased (Fig. 3.1b). Moreover, in the presence of TSC, the anodic peak of CT was slightly shifted to 0.54 V compared to that in the absence of TSC. The inset of Fig. 3.1c is the irreversible CV of 1 mM TSC, where the anodic wave is observed at 0.65 V. The increase in anodic peak height and decrease in cathodic peak height when TSC is present indicate that the electrogenerated *o*-quinone intermediate undergoes a chemical reaction with TSC later. This is an ECEC type of mechanism [181,182] and is represented in Scheme 3.1.

3.1.1 Effect of pH

The oxidation of CT in the presence of TSC is studied at various pH (Fig. 3.2). At each pH value, the CVs of CT and TSC are recorded, and the (I_{pa1}/I_{pa2}) ratios are calculated. I_{pa1} and I_{pa2} are the anodic

Electrochemistry of Dihydroxybenzene Compounds. DOI: http://dx.doi.org/10.1016/B978-0-12-813222-7.00003-6

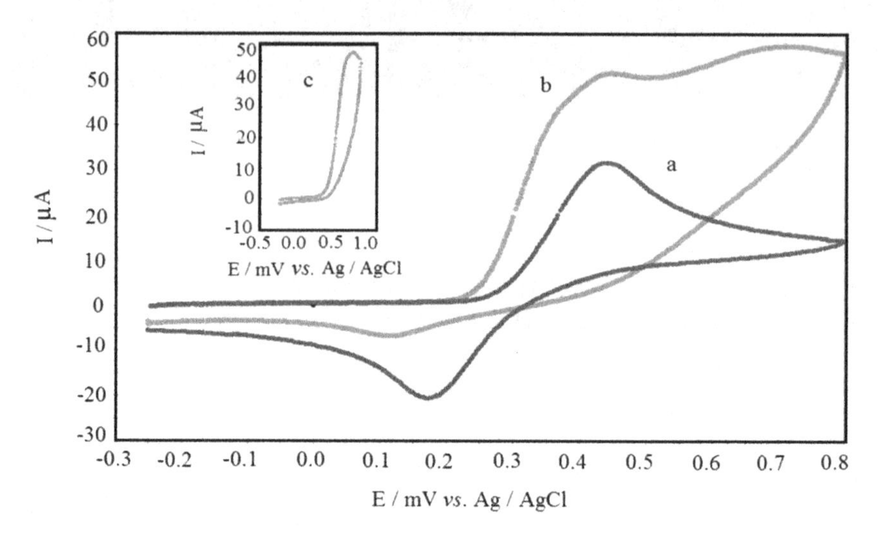

Figure 3.1 CVs of 1 mM CT in (a) the absence and (b) the presence of 1 mM TSC. Inset (c) is 1 mM TSC in the absence of CT, in acetate buffer (pH 5.3)–acetonitrile. Scan rate 50 mV s^{-1}.

Scheme 3.1 Electrochemical oxidation mechanism of CT in the presence of TSC [181,182].

peak currents of CT in the presence and absence of TSC, respectively. The CV at pH 3 and pH 4 exhibit an increase in I_{pa1}. The cyclic voltammetry of CT in the presence of TSC shows that there is a shift of the anodic peak potential (E_{pa}) in a positive direction and the cathodic peak potential (E_{pc}) began to disappear as pH values increase.

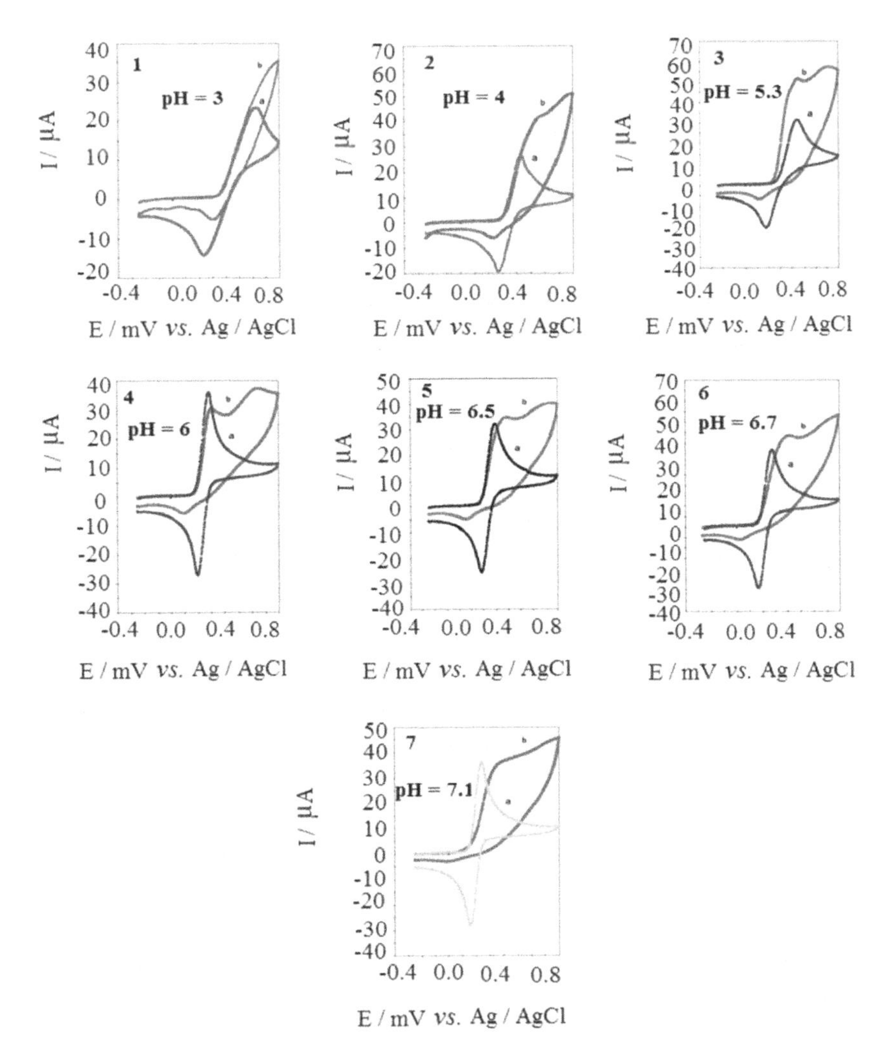

Figure 3.2 CVs of 1 mM CT at various pH in (a) the absence and (b) the presence of 1 mM TSC in water–acetonitrile (9:1) and 0.1 M acetate buffer at a GCE.

As the pH of the solution was lowered, the NH functionality was increasingly protonated, and hence, the nucleophilic character of the NH moiety diminished. Therefore, in acidic solutions, the reactivity of CT in the presence of TSC has increased with the increase in pH values [273]. The result revealed that the peak current ratio (I_{pa1}/I_{pa2}) increased with increasing pH up to pH 5.3 (Fig. 3.3). As a result, a decrease in the rate of dimerization with an increase in the rate of coupling is observed. The decrease in current–peak ratio after pH 5.3 could be related to a decline in the rate of the coupling reaction

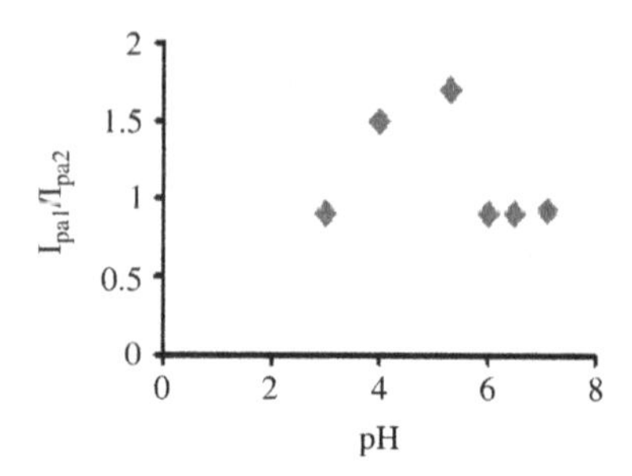

Figure 3.3 Plot of (I_{pa1}/I_{pa2}) versus pH.

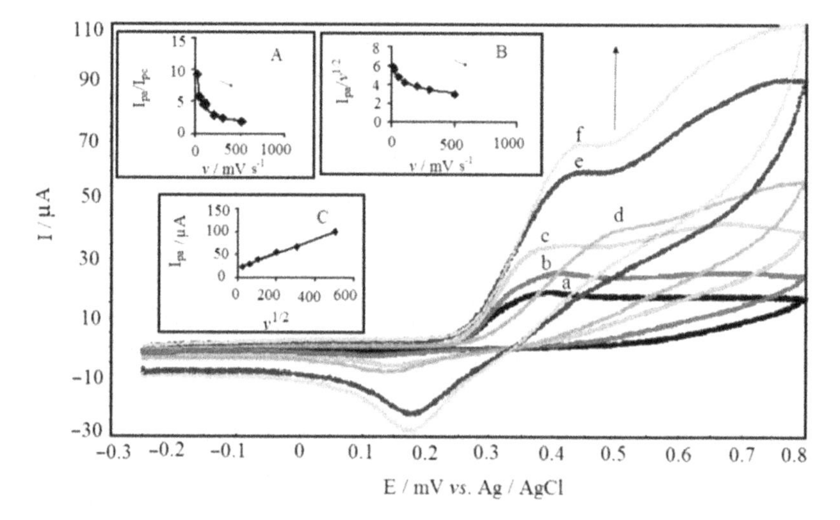

Figure 3.4 CVs of 1 mM CT in the presence of 1 mM TSC in water–acetonitrile (9:1) and 0.1 M acetate buffer (pH 5.3) at a GCE and at scan rates of (a) 20, (b) 50, (c) 100, (d) 200, (e) 300, and (f) 500 mV s^{-1}; Inset (A) variation of peak current ratio versus scan rate, (B) variation of peak current function versus scan rate, and (C) variation of anodic peak current versus square root of scan rate.

between CT and *o*-quinone. However, at pH higher than 7, the peaks could be ill-defined. This shows that the best pH for reaction between TSC and *o*-quinone is pH 5.3. Therefore, it is used for further studies.

3.1.2 Effect of Scan Rate

The CV of CT and TSC were scanned at different scan rates between 20 and 500 mV s^{-1} (Fig. 3.4). It appears that I_{pa} increases with the increase in scan rates. The plot of peak current ratio (I_{pa}/I_{pc}) versus scan rate confirms the reactivity of quinone with TSC through the increase in the height of I_{pc} at higher scan rates (Fig. 3.4 inset A).

As shown in CV a, the maximum ratio of I_{pa}/I_{pc} is obtained at a scan rate of 20 mV s^{-1}. Since the difference between peak current ratio values at scan rates 20 and 50 mV s^{-1} is small, the scan rate of 50 mV s^{-1} is then utilized due to the increase in the rate of analysis. Several authors have reported that the current function for the A_1 peak ($I_{pa}/\nu^{1/2}$), decreases with the increase in scan rate, and such behavior could be adopted as an ECEC mechanism (Fig. 3.4 inset B) [181,183,274]. However, oxidation currents (I_{pa}) linearly increased with the square root of the scan rate ($\nu^{1/2}$) suggesting that at sufficient over potentials, the reaction is diffusion controlled (Fig. 3.4 inset C).

3.2 CONTROLLED-POTENTIAL BULK ELECTROLYSIS FOR ELECTROORGANIC SYNTHESIS OF THE PRODUCT

For the determination of the number of "transferred electrons," the controlled-potential coulometry is performed in a water–acetonitrile mixture (9:1), and 0.1 M acetate buffer (pH 5.3) containing equimolar (0.5 mmol) amounts of both CT and TSC at 0.38 V versus Ag/AgCl (3 M NaCl). The process of electrolysis was monitored by CV. Fig. 3.5 shows the related progress of coulometry. The I_{pa} and I_{pc} decreases and finally disappears as the charge consumption becomes equals to

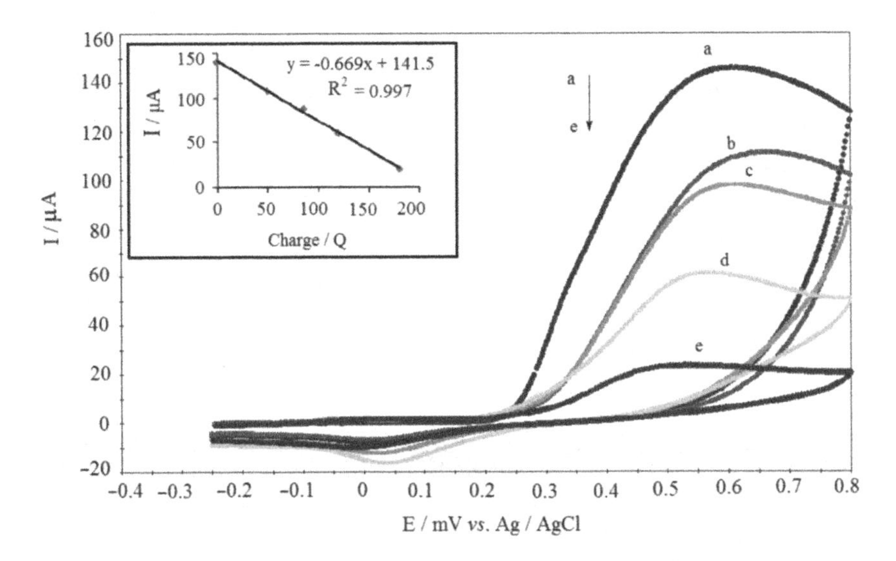

Figure 3.5 CVs of 0.5 mmol CT in the presence of 0.5 mmol TSC in H_2O–acetonitrile (9.1) and 0.1 M acetate buffer (pH 5.3) at a GCE during controlled-potential coulometry at 0.38 V versus Ag/AgCl (3 M NaCl), consumption Coulomb from a to e are: 0, 50, 86, 120, and 180 Coulomb, respectively. Inset: Variation of I_{pa} versus charge consumed, scan rate 50 mV s^{-1}.

$4e^-$ per molecule of CT (Scheme 3.1). Hence, the results obtained further emphasizes that an ECEC mechanism [181] has occurred during the electrooxidation of CT in the presence of TSC.

As results show, the formation of *o*-quinone (**2**) from the starting material (**1**) is evident, and one could expect that **2** would be attacked by the NH_2 or SH groups of TSC. The structure of **7** was characterized by using ^1H NMR, ^{13}C NMR, IR, and CHN.

3.3 CHARACTERIZATION OF 6,7-DIHYDROXY-1,2-DIHYDROBENZO[*E*] [1,2,4]-TRIAZINE-3(4*H*)-THIONE COMPOUND, 7

3.3.1 Elemental Analysis

The elemental analysis for **7** is presented in Table 3.1. The percentage of the elements found is in complete agreement with the theoretical values. The calculated molecular mass of the compound (**7**) was found to be equal to 197.21 g mol^{-1} and the isolated yield was 72%.

3.3.2 FTIR Analysis

The IR spectral data for the synthesized compound is in full agreement with the proposed structure. In general, the IR spectrum of compound **7** displays characteristic bands as depicted in Fig. 3.6. The C=C–H band is shifted to 3184 cm^{-1} from its original place in CT at 3400 cm^{-1}. The presence of aromatic moieties is easily recognizable by the presence of a sharp peak at 1498 cm^{-1} [275], which is attributable to the aromatic C=C stretching vibration [276,277]. However, this band usually appears at 1602 cm^{-1} in CT. An absorption band at 1163 cm^{-1} is assigned for C–N stretching vibration band in TSC. This band is shifted to1099 cm^{-1} in **7** [275,277]. The N–N stretching band appears at 920 cm^{-1} [278]. The two absorption bands at 3431 and 1660 cm^{-1} were assigned for the stretching and vibrations

Table 3.1 The Elemental Analysis of 7

Element	Composition (%)		
	Theoretical	Found	Δ
C	42.63	42.72	0.09
H	3.58	3.80	0.22
N	21.30	21.57	0.27

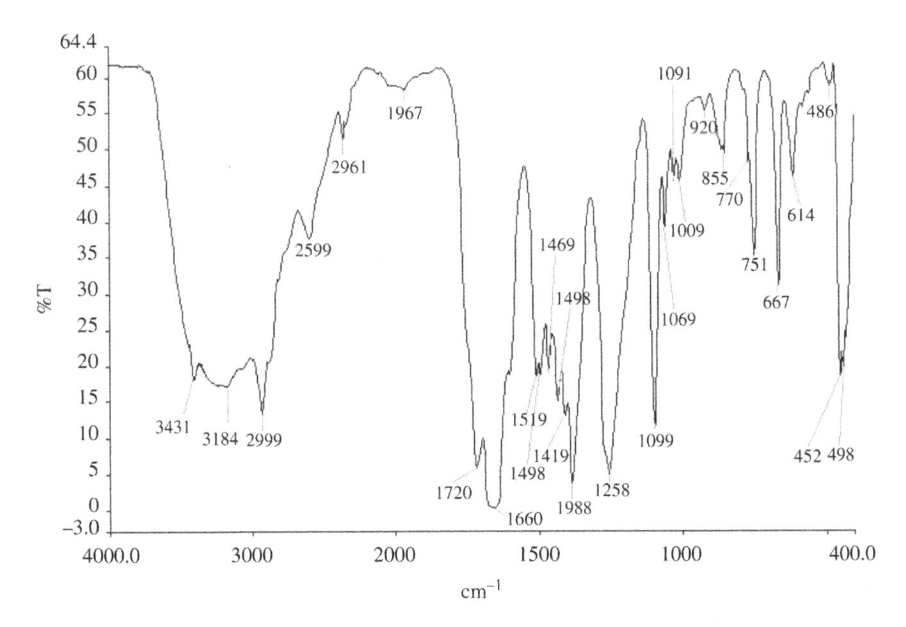

Figure 3.6 FT-IR spectrum of 7.

Table 3.2 Summary of the FT-IR Data Analysis of 7

Band	Frequency (cm^{-1})	Vibration
C = C–H	3184	Stretching
C = C (*aromatic*)	1498	Stretching
N–H	3431	Stretching
	1660	bending
C–N	1099	Stretching
N–N	920	Stretching
C = S	1513	Stretching

of N–H bond, respectively [275]. The band at 1513 cm^{-1} is related to C = S group [275]. Table 3.2 summarizes the IR results.

3.3.3 ^{1}H NMR and ^{13}C NMR Analysis

The NMR spectra were recorded in deuterated chloroform using TMS as internal standard. Chemical shifts are given in parts per million (δ scale) and the coupling constants are given in Hertz. The ^{1}H NMR spectrum of **7** confirmed the presence of eight proton signals (2.18, 4.01, 4.68, 5.19, 5.3, 5.85, and 6.02 ppm) (Fig. 3.7). The strong signal at 7.2 ppm was contributed by $CDCl_3$ solvent. The singlet signals at

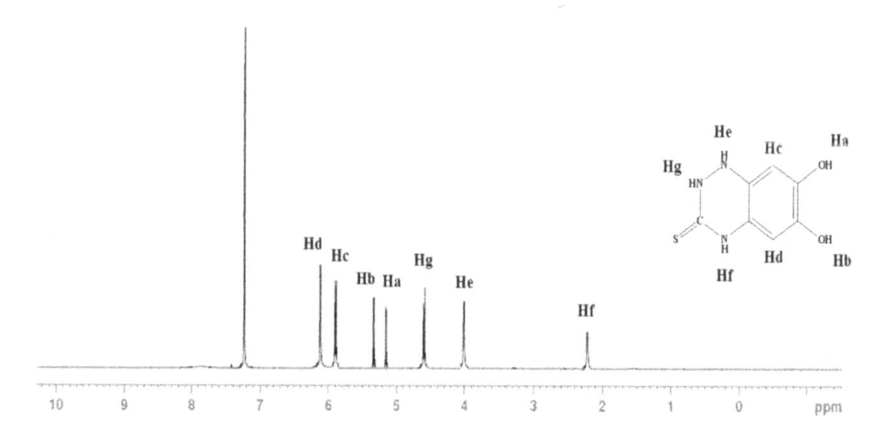

Figure 3.7 ^{1}H NMR spectrum of 7 in $CDCl_3$.

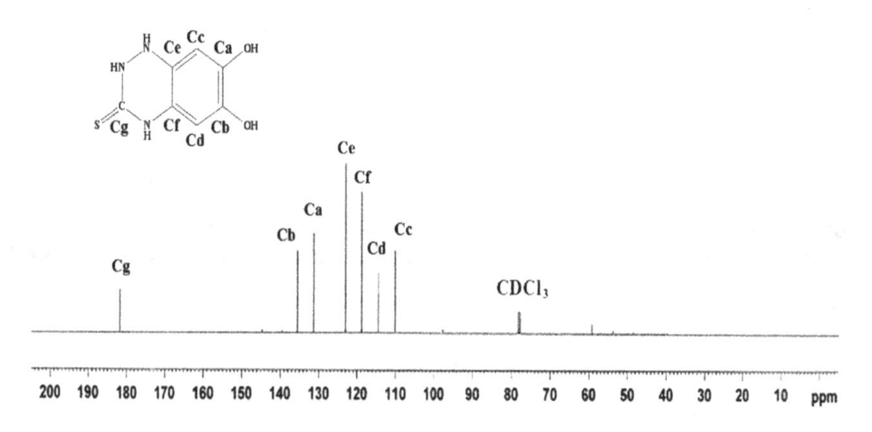

Figure 3.8 ^{13}C NMR spectrum of 7 in $CDCl_3$.

5.19 and 5.31 ppm are assignable to hydroxy groups, **Ha** and **Hb**, respectively. The signals at 5.85 and 6.02 ppm are attributable to aromatic hydrogens, **Hc** and **Hd**, respectively. The signals at 2.18 **(Hf)**, 4.01 **(He)**, and 4.68 **(Hg)** ppm indicated the presence of three amino groups (NH). These findings are in complete agreement with the electrosynthesis data obtained from 1,2,4-triazino[3,4-*b*]-1,3,4-thiadiazine derivatives [154].

The ^{13}C NMR spectrum of **7**, which depicted in Fig. 3.8, gives rise to eight signals at 77.91, 110.12, 115.04, 118.85, 122.89, 132.25, 138.81, and 182.85 ppm. The signal at 77.9 ppm was due to the solvent peak ($CDCl_3$). Peaks at 110.12 and 115.04 ppm are assignable to the CH (**Cc** and **Cd**, respectively). The two adjacent signals at 118.85 and

122.89 ppm are attributable to the aromatic carbons, **Cf** and **Ce**, respectively. Other signals at 132.25 (**Ca**) and 138.81 ppm (**Cb**) indicated the presence of C–OH groups. Further analysis using ^{13}C NMR displayed the presence of C = S carbon (**Cg**) at 182.85 ppm [275,279].

Therefore, both analytical and spectral data (^{1}H NMR, IR) of the studied compounds were in full agreement with the proposed structures. Accordingly, it is suggested that *o*-quinone (**2**) is probably attacked only by NH_2 group. The Michael addition reaction of **CT** to *o*-quinone (**2**), which provides to the intermediate **4**, is the fastest. The next oxidization is fast enough to form product **7** followed by a rapid intramolecular Michael-type addition. The final product (**7**) can undergo further oxidizations; however this is not the case in the present study. According to controlled-potential coulometry for oxidation of compound **7**, the number of the transferable electrons will increase from $4e^-$ to $6e^-$ per molecule. Our calculations show the presence of $4e^-$ per molecule (Fig. 3.5).

3.4 QUANTIFICATION OF TSC

The CV of 1×10^{-3} M CT at the glassy carbon electrode (GCE) in the presence of different concentrations of TSC under optimal conditions shows that the anodic peak of CT is linearly dependent on the concentration of TSC. The addition of TSC (10 to 850 μM) leads to an increase in the height of I_{pa} and decrease in I_{pc} (Fig. 3.9A). The successive decrease in I_{pc} can be ascribed to the fact that an increase in

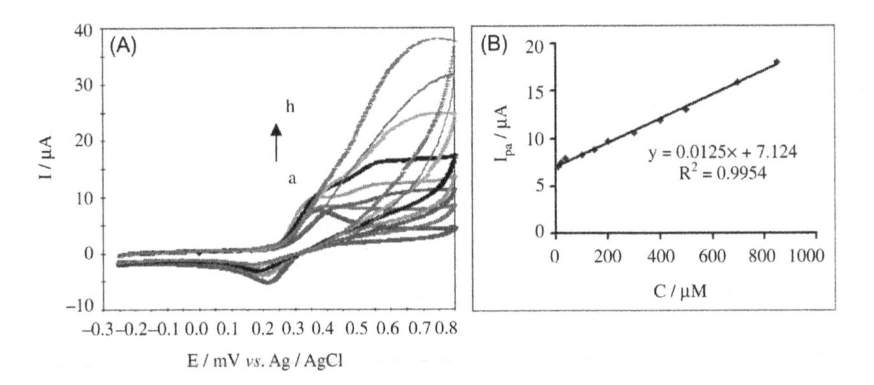

Figure 3.9 (A) CVs of 1×10^{-3} M CT in the presence of various concentrations of TSC (a-h; 1×10^{-5} to 8.5×10^{-4} M), scan rate: 50 mV s^{-1}. (B) Calibration plot for the determination of TSC in the presence of CT.

Table 3.3 F and T Tests for Determination of TSC

F Test		*T* Test	
Calculated ($n = 6$)	Tabulated (95%)	Calculated ($n = 6$)	Tabulated (95%)
2.56	5.05	1.34	2.57

concentration of TSC serves to scavenge the oxidized form of CT. However, on the reverse sweep there is only a little TSC left for electroreduction. Removal of the electrogenerated quinone through reaction with the NH_2 functionality of the TSC means that on the reverse sweep there is less quinone present to be electroreduced and, hence, the decrease in the height of the reduction peak. The increase in the magnitude of I_{pa} can be attributed solely to the reoxidation of CT-TSC adduct that have increased through the electrochemically initiated reaction previously detailed in Scheme 3.1 and Fig. 3.9B. The calibration plot obtained under the optimal conditions for a solution containing 1×10^{-3} M CT is linear in the range of 1×10^{-5} to 8.5×10^{-4} M with linear regression coefficient (R^2) of 0.995. The linear regression equations are $y = 0.0125x + 7.124$, where x is the concentration of TSC, and y is the peak current. The limit of detection (*dl*) of TSC 1.6×10^{-6} M is defined as $3S_B/m$, where S_B is the standard deviation of the blank and m is the slope of calibration graph. The relative standard deviation (RSD) of the method ($n = 6$) is 2%. The F and T tests were also carried out [280,281]. According to these two tests, there are no significant differences between the results obtained by either procedure at the 95% confidence level, indicating that the proposed method can be used for voltammetric determination of TSC (Table 3.3).

3.5 INTERFERENCE STUDIES

It is clear that an appropriate selection of the indicating species is important to ensure optimal performance of analysis. The electrochemical interaction between quinone and thiol has been shown to provide a highly selective method for the determination of different chemicals performed with a careful choice of indicators [282]. This action significantly minimizes the influence of common electroactive interferences. In order to assess possible analytical application of the method explained above, the effect of some organic and inorganic species of various concentrations is tested (Table 3.4).

Table 3.4 Interference of Some Organics and Inorganics Used for the Detection TSC	
Ions/Compounds Added	**Tolerance Limit (μM)**
K^+	450
Mg^{2+}, Ca^{2+}, Cu^{2+}, Co^{2+}, Ba^{2+}, Hg^{2+}, Pb^{2+}	nd
Ag^+	180
Acetone, methanol, ethanol	nd
Aniline	200
Thiourea	300
Urea, Ethylthiosemicarbazide	nd
2,4-Dinitrophenylhydrazine	nd
nd = not detected.	

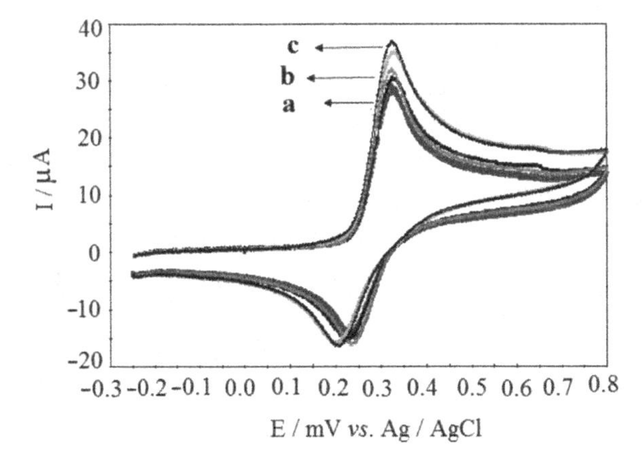

Figure 3.10 CVs of 1×10^{-3} M CT toward 2×10^{-4} M TSC, in (a) the absence and (b) the presence of 2×10^{-5}–4×10^{-4} M K^+, and (c) 4.5×10^{-5}–6×10^{-4} M K^+.

3.5.1 Effect of Foreign Ions

The interference due to several cations was studied in detail. Different amounts of ionic species were added to the 2×10^{-4} M TSC. The tolerance limit was taken as the amount required causing errors in the TSC recovery. As an example, the CV of 1×10^{-3} M CT in the presence of 2×10^{-4} M TSC was compared with the responses obtained in the presence of different concentrations of K^+ (2×10^{-5}–6×10^{-4} M), (Fig. 3.10). The CV shows that K^+ does not interfere, up to 4×10^{-4} M. However, subsequent additions of K^+ ($>4 \times 10^{-4}$ M) led to further increase with little discrimination in current between the various components. Though, increasing K^+ concentration to higher values, led to an increase in percentage error, no marked effect was observed for the CT response especially at low concentration of K^+.

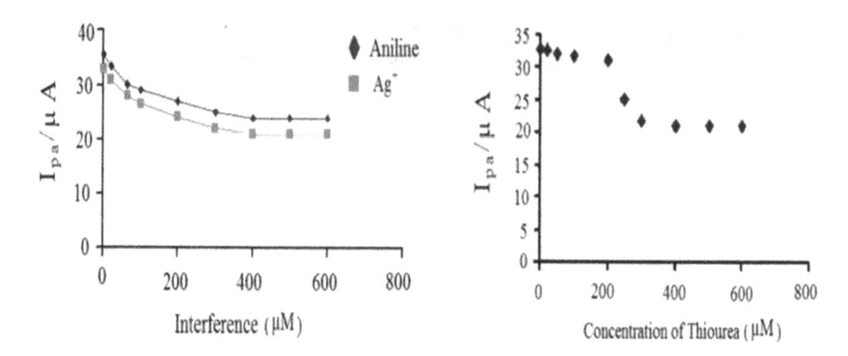

Figure 3.11 Influence of aniline, Ag^+, and thiourea on the analytical signal recorded for 1×10^{-3} M CT in the presence of 2×10^{-4} M TSC.

3.5.2 Effect of Organic Solvents and Some Organic and Inorganic Compounds

Organic solvents such as acetone, ethanol, and methanol may influence the response of CT in the presence of 2×10^{-4} M TSC. Results showed that these solvents do not interfere even in large amounts. The effect of some organic compounds, such as amines, semicarbazide derivatives, hydrazine derivatives, urea, and thiourea, might interfere with the reaction between quinone moiety, and TSC has also been investigated. Table 3.4 shows that the presence of aniline and thiourea lead to errors in excess of 100% which clearly oppose the responses observed for inorganic ions and organic solvents (Fig. 3.11). A slight depression in the CT/thiol signal by these organics may be attributed, at least in part, by surface passivation effects [282]. Deterioration in the electrode response is not a result of interaction between either the CT or CT-thiol adduct with the interferent, because such processes would have resulted in an enhancement in the current response. Reduction of the electrooxidized CT moieties by either interferent would have resulted in their immediate reoxidation at the electrode and hence an increased oxidation current. Fig. 3.11 shows the influence of aniline, Ag^+ and thiourea on the analytical signal recorded for CT in the presence of 2×10^{-4} M TSC. Other organic compounds and solvents studied do not interfere on the determination of TSC (Table 3.4).

3.6 APPLICATIONS

To confirm the usefulness of the suggested method, TSC is determined in tap water, lake water, and propranolol tablets spiked with

Table 3.5 Recoveries of TSC in Real Samples ($n = 5$)

Sample	RSD (%)	Recovery (%)
Tap water	2.8	93.3
Lake water	2.8	92.0
Propranolol tablet	2.2	94.2

TSC. The lake water sample was first filtered prior to the application. The pH of filtrate is 7.3. As mentioned in Section 2.5, the electrochemical measurements are conducted in two equal quantities of tap water and lake water samples at pH 5.3 (acetate buffer, 0.1 M) and acetonitrile as a cosolvent, containing a total of 2×10^{-4} M TSC. The electrochemical measurements for propranolol tablets are conducted in equal quantities of acetate buffer (0.1 M) pH 5.3 and acetonitrile containing a total of 2×10^{-4} M TSC and 8 mg propranolol that is soluble in distilled water. The recovery was obtained by measuring CV on the response of propranolol tablets for the CT reduction wave with subsequent additions of TSC. Similar behavior is observed for TSC in the presence of water samples. The RSD of this method, based on five replicates, is presented in Table 3.5. These findings indicate that the method is appropriate for the analysis of TSC in real samples.

3.7 ELECTROCHEMICAL CHARACTERIZATION OF P4VP/MWCNT–GCE

3.7.1 Surface Morphology Studies

The FESEM images of multiwalled carbon nanotubes (MWCNT), poly(4-vinylpyridine) (P4VP), and P4VP/MWCNT are shown in Fig. 3.12. The typical morphology of the well-aligned MWCNT and microspheres of P4VP are shown in Fig. 3.12A and B, respectively. The FESEM image of P4VP/MWCNT–GCE indicates that the MWCNT is densely covered by P4VP (Fig. 3.12C and D). The EFTEM images also confirm these findings (Fig. 3.12E and F). The P4VP in the nanocomposite appear to be distributed homogeneously throughout the material and are adherent to the CNT surface. The P4VP appear as dark spots embedded in the MWCNT matrix. These clearly reveal the coexistence of P4VP along with MWCNT on the GCE.

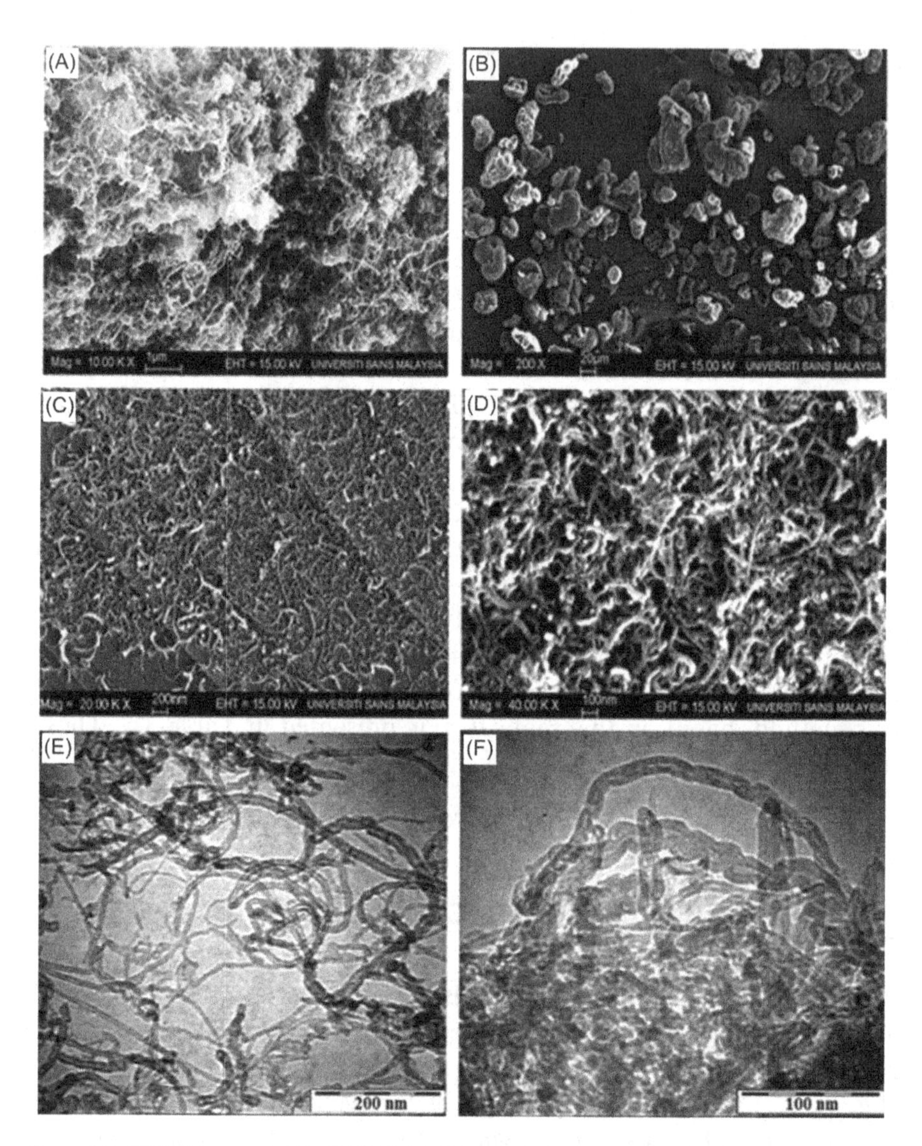

Figure 3.12 FESEM images of (A) MWCNT (10,000× magnification), (B) P4VP (200× magnification), and P4VP/MWCNT in (C) low-magnification (20,000× magnification) and (D) high-magnification (40,000× magnification), EFTEM images of P4VP/MWCNT in (E) low-magnification and (F) high-magnification.

3.7.2 CV Studies

The cyclic voltammetry of 5 mM $[Fe(CN)_6]^{3-/4-}$ in 0.5 M KCl at the surface of the bare GCE, MWCNT–GCE, and P4VP/MWCNT–GCE were carried out over a potential range of −0.25 to 0.8 V. Fig. 3.13 shows that the peak current is more noticeable for the P4VP/MWCNT–GCE indicating a greater electroactive surface area.

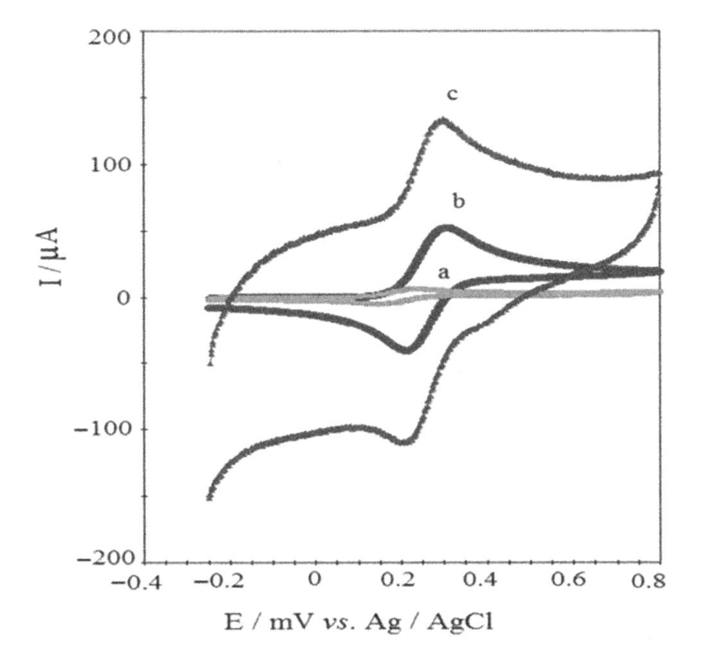

Figure 3.13 CVs of 5 mM $[Fe(CN)_6]^{3-/4-}$ in 0.5 M KCL at (a) bare GCE, (b) MWCNT–GCE, and (c) P4VP/MWCNT–GCE, scan rate 100 mV s^{-1}.

3.7.3 EIS Studies

The AC impedance spectra of bare GCE, MWCNT–GCE, and P4VP/MWCNT–GCE were also looked into. Fig. 3.14A shows Nyquist plots of 5 mM $[Fe(CN)_6]^{3-/-4}$ at these electrodes within the frequency range of 100 kHz to 100 MHz at the formal potential of the redox probe. The almost straight line plots obtained implies the low charge transfer resistance of the redox probe [283,284]. At high frequencies, MWCNT–GCE (plot a) experiences a sharp increase of Z_{im} and behaves closer to an ideal capacitor [285]. The slight variation from the ideal capacitive behavior could be due to the pore size distribution of MWCNTs [286–288]. The P4VP/MWCNT–GCE (plot b) demonstrates a deviation from an ideal capacitor and shows a much lower impedance than the pure MWNT–GCE (plot a), P4VP/GCE (plot c) and bare GCE (plot d). The presence of a small and depressed semicircle in plots a and b (more pronounced for the bare GCE than the P4VP/MWCNT–GCE at higher frequencies) could be due to the charge-transfer resistance (R_{ct}) of the electrodes. The extended semicircle loop in the high to medium frequency region displayed by the P4VP/GCE could indicate the high R_{ct} and Warburg element (Z_w) of diffusion limitations as compared to the other

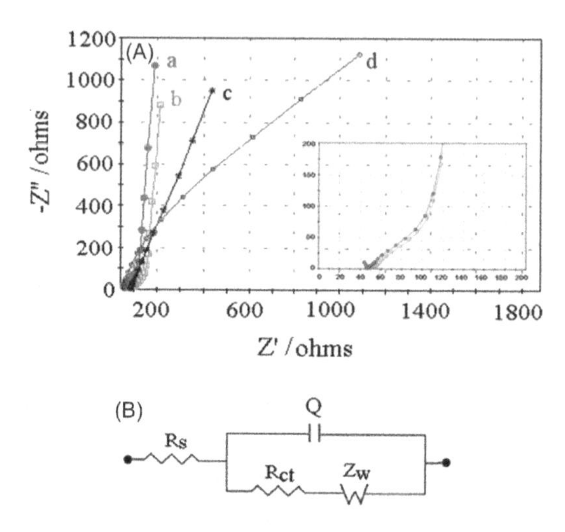

Figure 3.14 (A) Nyquist plots of (a) MWCNT–GCE, (b) P4VP/MWCNT–GCE, (c) P4VP/GCE, and (d) bare GCE in 0.5 M KCl supporting electrolyte containing 5 mM $[Fe(CN)_6]^{3-/4-}$. (B) Eletrochemical equivalent circuit for the system.

two electrodes. Fig. 3.14B is the modified equivalent circuit from data fitted in the high frequency region. In this circuit, R_s and Q represented solution resistance and a constant phase element corresponding to the double-layer capacitance. It can be summarized that the sequence of R_{ct} is in the order of P4VP/MWCNT–GCE (23 Ω) < MWNT–GCE (25 Ω) < P4VP/GCE (30 Ω) < GCE (1640 Ω). As the P4VP/MWCNT–GCE has a lower R_{ct} implying that the interfacial resistance has decreased and the conductivity and the electron transfer process improved when P4VP is present in the nanocomposite-modified electrode as compared to the GCE and MWNT–GCE. Therefore, it can be concluded that the nanocomposite P4VP/MWCNT–GCE benefitted from the synergistic effect of P4VP and MWNT as well as P4VP being an electron mediator in the electron transfer process.

3.8 ELECTROCHEMICAL CHARACTERIZATION OF P4VP/GR–GCE

3.8.1 CV Studies

The cyclic voltammetry of 5 mM $[Fe(CN)_6]^{3-/4-}$ in 0.5 M KCl at the surface of the bare GCE, GR–GCE, and P4VP/GR–GCE were carried out over a potential range of −0.25 to 0.8 V. Similar to Section 3.7.2, Fig. 3.15 also shows that peak current is more noticeable for the P4VP/GR–GCE indicating a greater electroactive surface area.

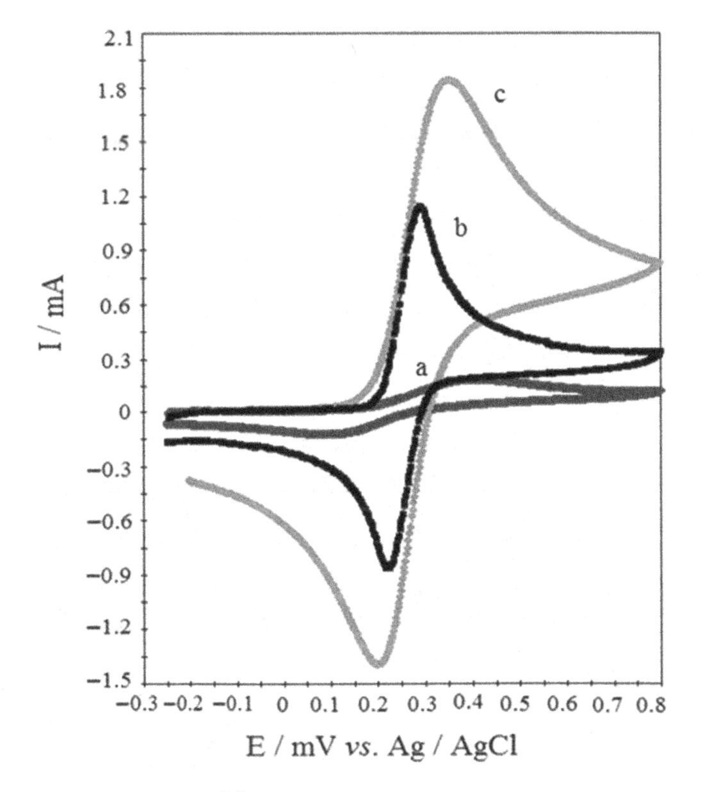

Figure 3.15 CV of 5 mM $[Fe(CN)_6]^{3-/4-}$ in 0.5 M KCl at (a) bare GCE, (b) GR−GCE, and (c) P4VP/GR−GCE, scan rate 100 mV s^{-1}.

3.8.2 Surface Morphology Studies

The SEM images of P4VP/GR (Fig. 3.16C and D) show that microspheres of P4VP are well distributed between or on the paper like GR nanosheets, which indicates GR is densely covered by P4VP. The TEM analyses in Fig. 3.16E and F also confirm these findings. The P4VP appears as dark spots homogeneously distributed throughout the composite material and adherent to the GR surface and even embedded in the entangled transparent GR sheets. These results clearly reveal that the P4VP along with GR existed as a composite and demonstrated the formation of dense films in the GCE.

3.8.3 Electrochemical Impedance of P4VP/GR−GCE

Electrochemical impedance (EIS) is widely used to study the interfacial properties of solid electrodes and solution bulk for information such as electron transfers, electrode impedances, electric double layer, and the surface charge transfer resistance. The results are usually shown as a

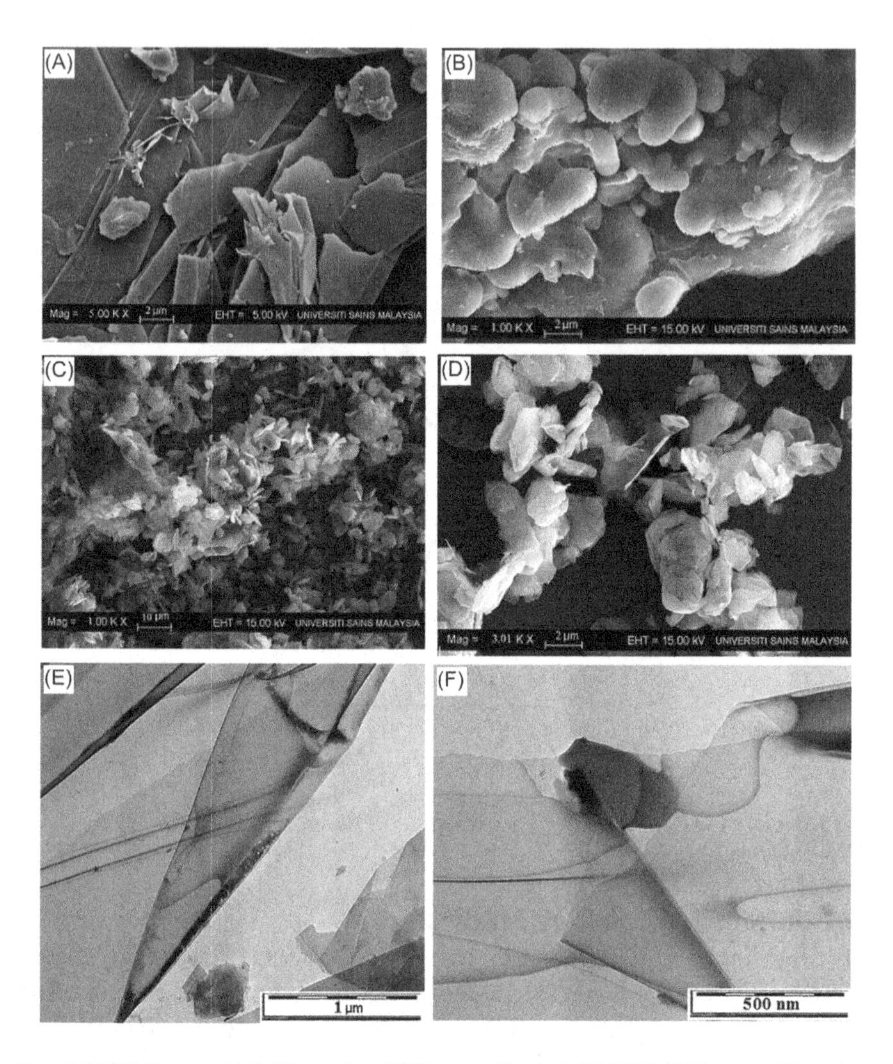

Figure 3.16 SEM images of (A) GR nanosheet (5000× magnification) (B) P4VP (1000× magnification) and P4VP/GR in (C) low-magnification (1000× magnification), and (D) high-magnification (3000× magnification), EF-TEM images of P4VP/GR in (E) low-magnification and (F) high-magnification.

semicircular part at higher frequencies corresponds to the kinetic control, and a linear part at lower frequencies corresponds to the mass transfer control [238,283]. The diameter of the semicircle represents the R_{ct} at the electrode surface while the linear part represents Warburg impedance (Z_w). The Nyquist plots of 5 mM $[Fe(CN)_6]^{3-/4-}$ at bare GCE, P4VP/GCE, GR−GCE, and P4VP/GR−GCE within the frequency range of 0.1 Hz−10 kHz are shown in Fig. 3.17. The figure revealed very small and depressed semicircle in plots a, b, c, and

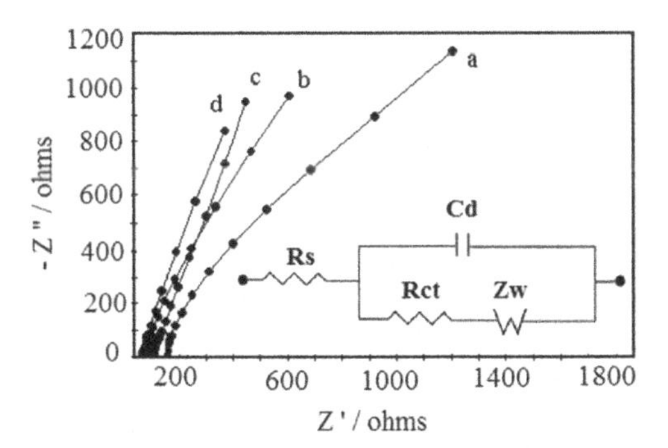

Figure 3.17 Nyquist plots of (a) bare GCE, (b) P4VP/GCE, (c) GR–GCE, and (d) P4VP/GR–GCE in 0.5 M KCl electrolyte solution containing 5 mM $[Fe(CN)_6]^{3-/4-}$. Inset: Eletrochemical equivalent circuit for the system.

d (more pronounced for the bare GCE than others). The presence of depressed semicircle in the bare GCE (plot a) over the high frequency range can be attributed to the poor electrical conductivity of this electrode compared to other electrodes. In addition, the straight line in the low frequency region of the bare GCE and the P4VP/GCE (plot b) indicate greater variations in the frequency dependence of ion diffusion and increased obstruction of ion movement in electrolyte [289]. On the other hand, the P4VP/GR–GCE (plot d) has little higher interfacial R_{ct} than GR (plot c). However, both of them are very low as compared to the P4VP/GCE and the bare GCE due to the good conductivity of GR [138]. By fitting the data to verify modified equivalent circuit (inset of Fig. 3.17), the values of R_{ct} for different electrodes are in the order of GR–GCE (21 Ω) < P4VP/GR–GCE (22 Ω) < P4VP/GCE (26 Ω) < GCE (1640 Ω). Moreover, the neglected Warburg region on the Nyquist plot of P4VP/GR–GCE indicates short and equal diffusion path length of the ions in the interfacial region. This can be attributed to the homogenous morphology of the P4VP/GR–GCE composite into which the ions of electrolyte have access to the surface of the nanocomposite [138,289].

3.9 ELECTROCHEMISTRY OF HQ AND CT ON THE P4VP/MWCNT–GCE

Fig. 3.18 shows CVs obtained for the bare GCE, MWCNT, and P4VP/MWCNT–GCE electrodes in the supporting electrolyte (buffer pH 2.5). A bare GCE (Fig. 3.18a) in the absence of CT and hydroquinone (HQ) has exhibited only a baseline and poor electrochemical

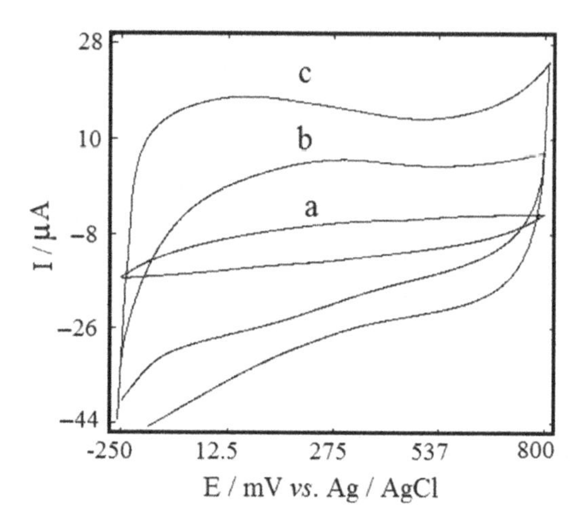

Figure 3.18 CVs obtained for the (a) bare GCE, (b) MWCNT−GCE, and (c) P4VP/MWCNT−GCE in 0.1 M sodium sulfate buffer solution (pH 2.5) at scan rate 20 mV s^{-1}.

response. However, the peak current is more noticeable for MWNT−GCE (Fig. 3.18b) and especially for the P4VP/MWCNT−GCE which indicate a much greater electroactive surface area than other electrodes (Fig. 3.18c).

The electrochemical responses of 2 μM CT and HQ in 0.1 M sodium sulfate buffer solution (pH 2.5) at three different electrodes have been studied by using CV (Fig. 3.19). The broad redox couple peaks at the bare GCE (Fig. 3.19a) and MWCNT (Fig. 3.19b) for CT and HQ could indicate a sluggish rate of electron transfer. However, the P4VP/MWCNT−GCE for detection of CT displays a pair of well-defined redox peaks with E_{pa} at 349 mV and E_{pc} at 301 mV. The ΔE_p of 48 mV indicates a favorable reversible electrode process (Fig. 3.19A c). An almost similar result was obtained for HQ (Fig. 3.19B c). The redox peaks at E_{pa} (232 mV) and E_{pc} (167 mV) with ΔE_p of 65 mV also indicate a favorable quasireversible electrode process. While at MWCNT−GCE, ΔE_p values for CT and HQ are 100 and 110 mV, respectively. Hence, smaller ΔE_p for CT and HQ at the P4VP/MWCNT−GCE indicates the high reversibility of the electrode process as a result of faster kinetics of electron transfer [138] when P4VP is present in the nanocomposite-modified electrode as compared to the MWCNT−GCE and bare GCE. Moreover, based on EIS studies the R_{ct} indicates that the kinetics of charge transfer at the P4VP/MWCNT−GCE is very favorable. The synergistic effect of MWCNT

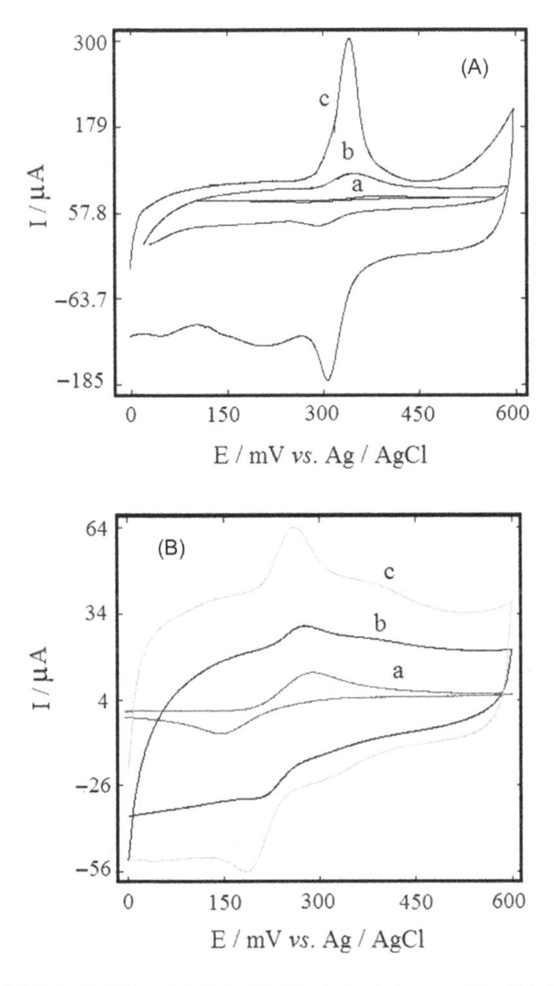

Figure 3.19 CVs of (A) 2 μM CT and (B) 2 μM HQ at the (a) bare GCE, (b) MWCNT−GCE, and (c) P4VP/MWCNT−GCE in 0.1 M sodium sulfate buffer solution (pH 2.5) at scan rate 20 mV s^{-1}.

and P4VP in the P4VP/MWCNT nanocomposite has provided an effective microenvironment for facile redox processes of these diphenols. It is also obvious that the increase in peak currents is due to the electrocatalytic properties of the MWCNT and P4VP toward CT.

The voltammetric behaviors of HQ and CT mixture (2 μM each) are shown in Fig. 3.20. At a bare GCE (CV a), the electrochemical response only shows a baseline and at the MWCNT−GCE (CV b), it was still not possible to separate and detect the HQ and CT in the mixture despite the increase in anodic response. However, at P4VP/MWCNT−GCE a pair of well-defined redox peaks (peak 1 and 3 in

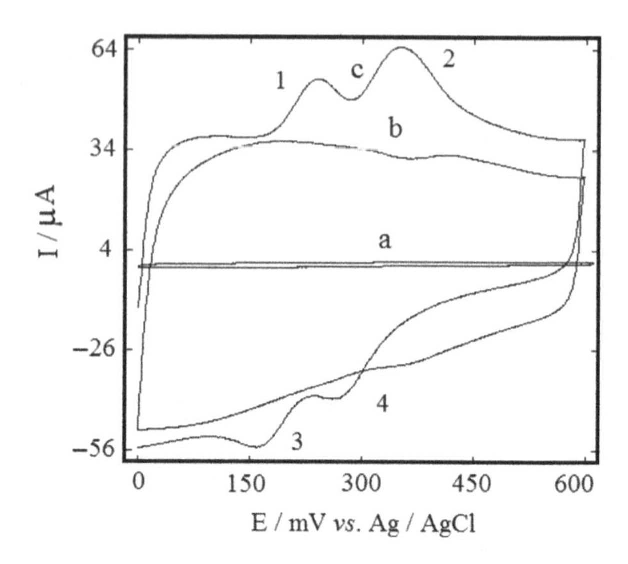

Figure 3.20 CVs of 2 μM HQ and 2 μM CT at the (a) bare GCE, (b) MWCNT−GCE, and (c) P4VP/MWCNT−GCE in 0.1 M sodium sulfate buffer solution (pH 2.5) at scan rate 20 mV s^{-1}.

Fig. 3.20) which corresponds to the redox of HQ appeared at 231 and 166 mV with a potential separation (ΔE) of 65 mV. Another pair of redox peaks (peaks 2 and 4) appeared at 348 and 286 mV with a ΔE of 62 mV, which is the redox of CT. Moreover, the mixture of HQ and CT at the P4VP/MWCNT−GCE (Fig. 3.20C) has shown two well defined redox peaks (peaks 1 and 2 in Fig. 3.20) with a ΔE_p of about 117 mV. More interestingly, the I_{pa} have also increased significantly. The oxidation current of HQ ($I_{pa} = 53$ μA) and CT ($I_{pa} = 64$ μA) on the P4VP/MWCNT−GCE is around 50 times higher than that on the bare GCE. Thus, the reaction mechanism to elucidate the electrode process of CT and HQ at P4VP/MWCNT−GCE is suggested in Scheme 3.2.

Scheme 3.2 shows that P4VP/MWCNT−GCE acts as an electron mediator to accelerate the electron transfer rate for oxidations of CT and HQ. It is also noticeable that the oxidation of DHB has produced quinone and vice versa [235]. Thus, the composite film with different mass ratios of MWCNT and P4VP (i.e., 4:2, 2:4, 1:1, 3:2, and 2:3) were fabricated. Then the effects of these ratios on the voltammetric responses of the nanocomposite electrodes for the detection of HQ and CT were investigated (Fig. 3.21). As we can see, a well-defined redox peak with a significant I_{pa} is obtained in the mass ratio of 4:2 of

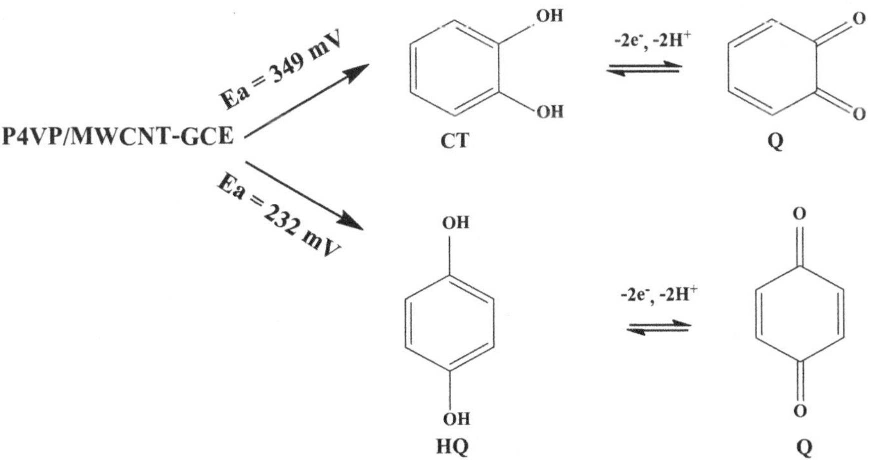

Scheme 3.2 The expected redox mechanism of HQ and CT at P4VP/MWCNT–GCE. Q is quinone.

MWCNT and P4VP. In this regard, the effects of pH and scan rate have been studied on the modified electrode with mass ratio of MWCNT and P4VP at 4 to 2.

3.9.1 Effects of Solution pH

The effect of solution pH on the electrochemical response of the P4VP/MWCNT–GCE toward 2 μM CT and HQ was investigated using CV. Variations of peak currents with respect to pH of the electrolyte in the pH range of 1.5 to 7 are shown (Fig. 3.22). It can be seen that the I_{pa} of CT increases with solution pH until the pH reaches 2.5. However, a small current was detected when the pH of the solution was either lower or higher than 2.5 which indicated that the P4VP/MWCNT–GCE is suffering from either a rapid or permanent loss of activity. The decrease in peak current after pH 2.5 could be related to a decline in the rate of the coupling reaction between CT and quinone [290]. A similar experiment was done for HQ. The results also show that at pH 2.5, the oxidation of HQ has a high electrochemical response.

3.9.2 Effect of Scan Rate

The scan rate dependence of the modified electrode for 2 μM CT and HQ was also investigated (Fig. 3.23). As the scan rate increases, the I_{pa} is increased and the E_{pa} is shifted slightly toward positive potentials. This could be due to increase in the electrocatalytic activity of

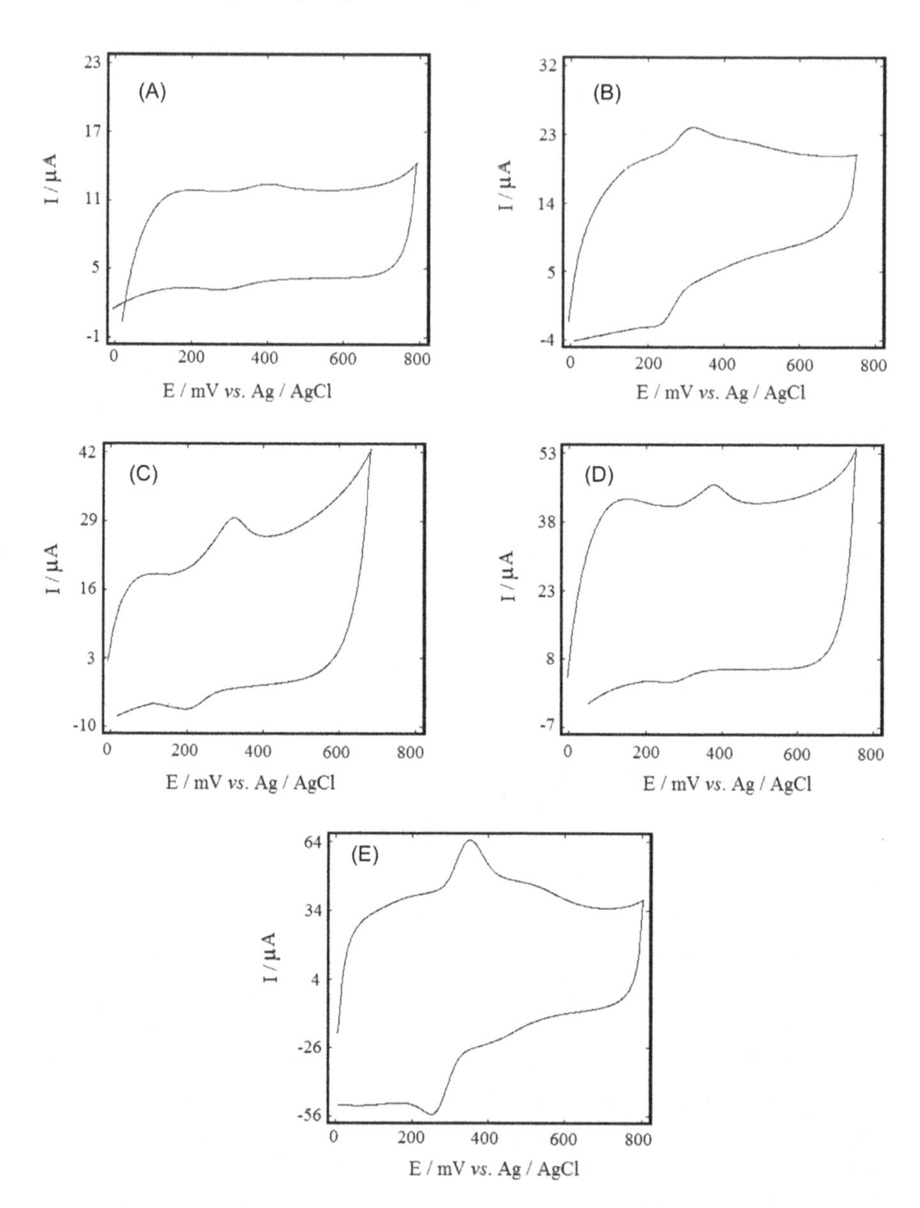

Figure 3.21 CVs of (A) 2 μM HQ at different mass ratio of MWCNT and P4VP (A) 2:4, (B) 2:3, (C) 1:1, (D) 3:2, and (E) 4:2 in 0.1 M sodium sulfate buffer (pH 2.5) at scan rate 20 mV s^{-1}.

the surface of P4VP/MWCNT–GCE on the oxidation of CT and HQ at scan rates higher than 100 mV s^{-1}. The I_{pa} has a linear relationship with the ($\nu^{1/2}$) suggesting that at the over potentials applied the reaction is diffusion controlled. This has been successfully carried out by obtaining the CV of HQ and CT in sodium sulfate buffer.

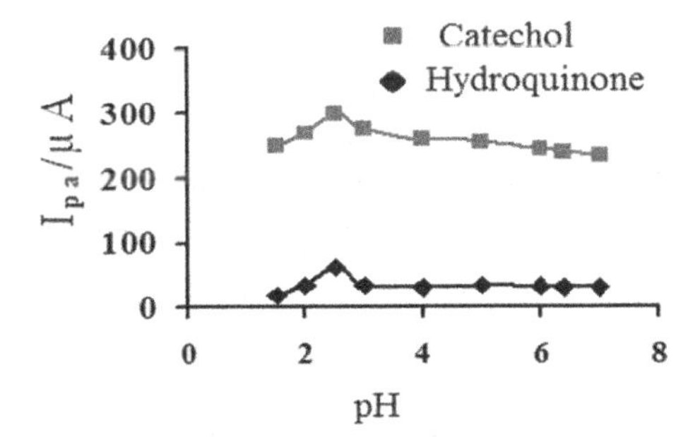

Figure 3.22 Effect of the pH on the anodic peak currents.

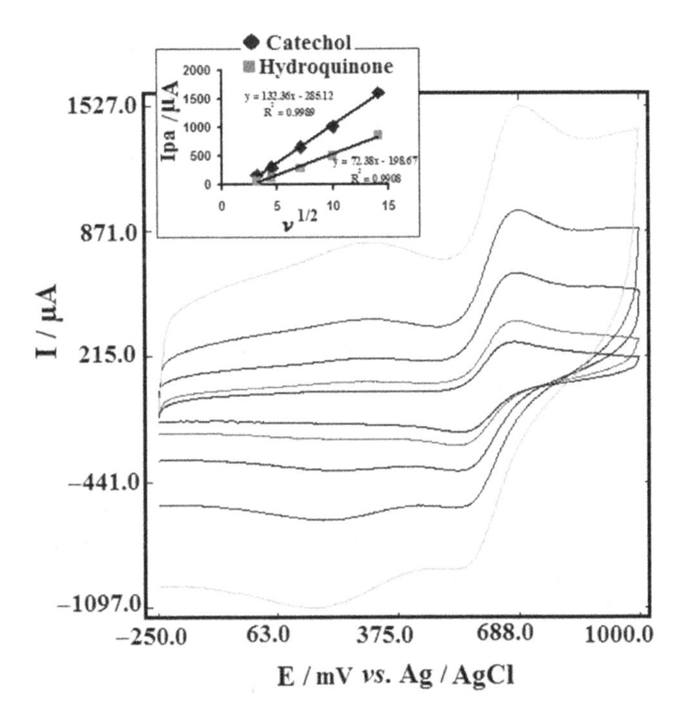

Figure 3.23 CVs of 2 µM CT in pH 2.5 sodium sulfate buffer solution at P4VP/MWCNT−GCE at scan rates (a) 10, (b) 20, (c) 50, (d) 100, and (e) 200 mV s^{-1}. Inset: Linear relationship of anodic peak current for 2 µM CT and HQ versus square root of scan rate.

3.10 DETERMINATION OF CT AND HQ USING DPV

DPV is used for the quantitative determination of CT and HQ (calibration plots) at P4VP/MWCNT−GCE. This is due to its better current sensitivity and resolution than CV. The DPV of different concentrations of CT and HQ in 0.1 M pH 2.5 buffer (sodium sulfate buffer) with applied potentials −0.25 to 0.8 V, step potential 2.0 mV, modulation amplitude 50 mV, and scan rate 10 mV s^{-1} (Fig. 3.24). The I_{pa} is linearly proportional to the concentration of CT (Fig. 3.24A) and HQ (Fig. 3.24B) in the range of 0.2 to 6 μM and 0.2 to 8 μM, respectively. Accordingly, two linear equations I_{pa} (μA) = 94.52 [CT] (μM) + 92.39 and I_{pa} (μA) = 10.21[HQ] (μM) + 37.36 with the respective linear regression coefficients of (R^2) 0.999 and 0.995 were obtained. The detection limits of HQ and CT are 15 nM and 2.6 nM, respectively. Fig. 3.24C shows the results of P4VP/MWCNT–GCE on simultaneous determination of HQ and CT at various concentrations. The DPV results indicate well-separated peaks at potentials 231 and 348 mV which correspond to the HQ and CT, respectively. This confirms that the modified electrode was suitable for the simultaneous detection of CT and HQ in a mixed system of dihydroxybenzenes. The method is also simple, rapid, and accurate. Table 3.6 shows the comparison between the results for the determination of dihydroxybenzene isomers. The limit of detections of HQ and CT at P4VP/MWCNT–GCE in this study indicate that the values obtained are far better than those reported.

3.11 APPLICATION TO REAL SAMPLE ANALYSIS

The developed electrode was tested for the simultaneous determination of CT and HQ in water samples (tap water and lake water). The lake water sample was filtered and the pH of filtrate was maintained at 7.3 [290]. Since the amount of HQ and CT are unknown in water samples, the standard addition method is applied by spiking known concentrations of the analytes into the samples. The recovery was obtained by measuring DPV responses for samples in which CT and HQ were added. The RSD of this method, based on five replicates ($n = 5$), is presented in Tables 3.7 and 3.8. Satisfactory recoveries percentages of CT and HQ at P4VP/MWCNT–GCE in the range of 0.2 to 8 μM show that this method is effective and reliable. These findings indicate that the method is suitable for the effective and sensitive analysis of CT and HQ.

3.12 INTERFERENCE STUDIES

Phenol and nitrophenol are the traditional interferences in the determination of 2 μM CT and HQ. The CVs of nitrophenol and phenol at different concentrations from 0.1 to 10 μM on the surface of the

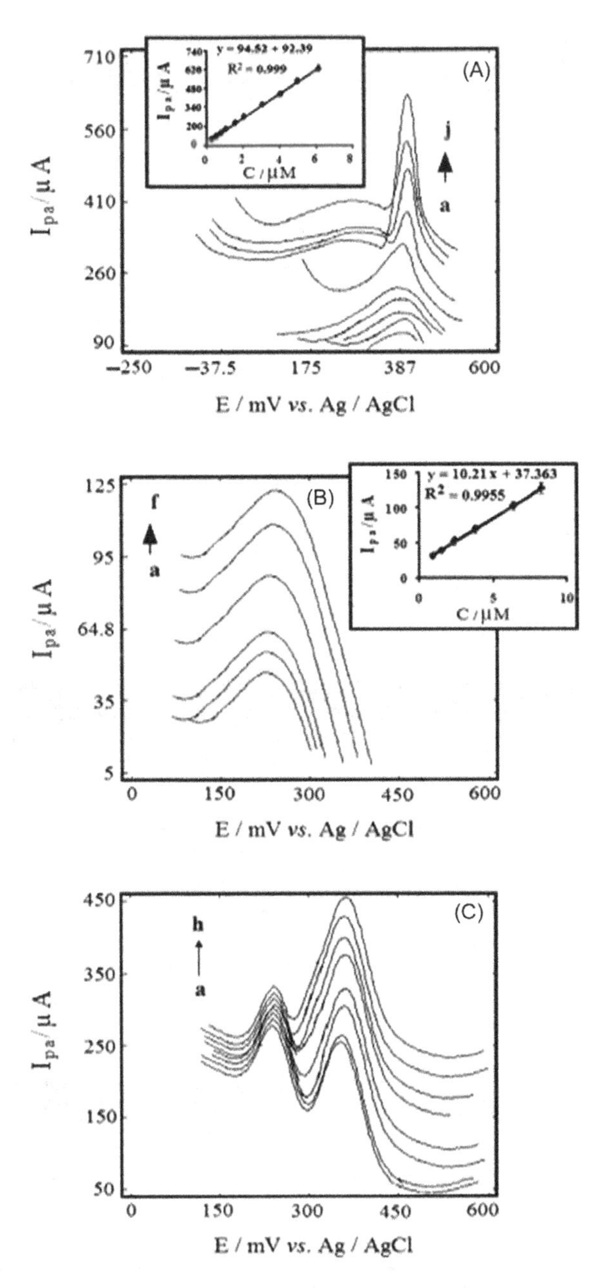

Figure 3.24 DPVs at P4VP/MWCNT–GCE of (A) (a) 0.2, (b) 0.5, (c) 0.7, (d) 1, (e) 1.5, (f) 2, (g) 3, (h) 4, (i) 5, and (j) 6 µM CT; (B) (a) 0.2, (b) 1, (c) 2, (d) 4, (e) 6, and (f) 8 µM HQ; and (C) (a) 0.2, (b) 0.5, (c) 1, (d) 2, (e) 3, (f) 4, (g) 5, and (h) 6 µM HQ and CT in sodium sulfate buffer (pH 2.5).

Table 3.6 Responses of Some of HQ and CT Sensors Constructed from Various Electrode Materials

Electrode Material	Technique	Detection Limit	Linear Range (μM)	Ref.
Graphene	DPV	0.8 μM	1–10 and 10–80	[229]
MWNT-IL-Gel/GCE	DPV	0.067 μM HQ	0.2–35	[235]
		0.06 μM CT	0.18–35	
Poly(thionine)-modified GCE	DPV	30 nM HQ	1–120	[233]
		25 nM CT		
p-Phenylalanine-modified GCE	DPV	1.0 μM HQ	10–140	[236]
		0.7 μM CT		
Electrochemically activated GCE	SWVs	0.018 μM HQ	1–100	[238]
		0.032 μM CT	2–100	
Graphene-modified GCE	DPV	0.015 μM HQ	1–50	[291]
		0.01 μM CT		
P4VP/MWCNT–GCE	DPV	15 nM HQ	0. 2–8	Present study
		2.6 nM CT	0.2–6	

Table 3.7 Recoveries of CT and HQ in Tap Water ($n = 5$)

Sample	Tap Water Containing CT	HQ Added (μM)	CT Found (μM)	Recovery (%)	RSD (%)
1	1.49	0	1.35	90.6	2.2
2	1.49	10	11.36	98.6	1.3
3	1.49	20	21.56	100.3	1.8
Sample	**Tap Water Containing HQ**	**CT Added (μM)**	**HQ Found (μM)**	**Recovery (%)**	**RSD (%)**
1	19.04	0	19.20	100.8	1.0
2	19.04	10	29.00	99.8	1.7
3	19.04	20	38.6	98.8	1.0

Table 3.8 Recoveries of CT and HQ in Lake Water ($n = 5$)

Sample	Lake Water Containing CT	HQ Added (μM)	CT Found (μM)	Recovery (%)	RSD (%)
1	1.44	0	1.43	99.3	2.7
2	1.44	10	11.46	100.1	1.3
3	1.44	20	21.06	98.2	2.8
Sample	**Lake Water Containing HQ**	**CT Added (μM)**	**HQ Found (μM)**	**Recovery (%)**	**RSD (%)**
1	19.65	0	19.46	99.0	1.0
2	19.65	10	29.16	98.3	1.2
3	19.65	20	39.53	99.6	1.0

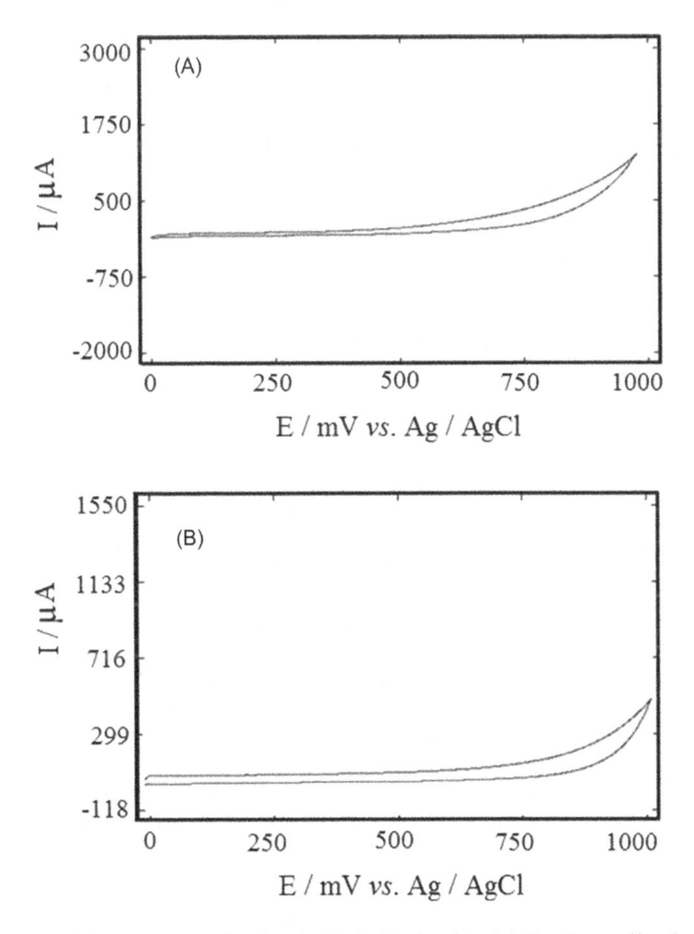

Figure 3.25 CVs of (A) 2 μM nitrophenol and (B) 2 μM phenol in 0.1 M sodium sulfate buffer solution (pH 2.5) at P4VP/MWCNT–GCE with a scan rate 20 mV s^{-1}.

modified electrode in the potential range of 0 to 1 V show that these compounds do not interfere even in large amounts (Fig. 3.25). Consequently, the coexistence of these phenolic compounds do not interfere with the quantitative detection of HQ and CT. In addition, no significant interference from common cations such as Ca^{2+}, Mg^{2+}, Mn^{2+}, Zn^{2+}, Fe^{3+}, Cu^{2+}, and Co^{2+} was observed.

3.13 REPRODUCIBILITY AND STABILITY OF P4VP/MWCNT–GCE

The regeneration of the surface of electrode was examined by cyclic voltammetric studies of seven different electrodes constructed by the same procedure. The calculated RSD of 3.7% ($n = 7$) for various parameters was accepted as the criteria for a satisfactory surface reproducibility which is practically the same as that expected for the renewed electrode. In addition, the long-term stability of the modified electrode was also evaluated. After the modified electrode was stored for 2 months, it retained 99% of its initial current sensitivity. It is obvious that the modified electrode has excellent stability and reproducibility.

3.14 ELECTROCHEMISTRY OF DIPHENOLS ON THE P4VP/GR–GCE

Fig. 3.26 shows CVs obtained for the bare GCE, GR–GCE, and P4VP/GR–GCE electrodes in supporting electrolyte (buffer pH 2.5). A bare GCE in the absence of CT and HQ has only baseline and poor electrochemical response (Fig. 3.26a). The peak currents of MWNT–GCE (Fig. 3.26b) and P4VP/GR–GCE (Fig. 3.26c) are increased, indicating greater electroactive surface area.

The electrochemical responses of CT and HQ in 0.1 mol L^{-1} sodium sulfate buffer (pH 2.5) at three different electrodes have been studied using CV. The results show a broad and oxidation peak of CT (Fig. 3.27A) and HQ (Fig. 3.27B) at the bare GCE (CV a) and GR–GCE (CV b) indicating that the rate of electron transfer is a bit sluggish. However, the P4VP/GR–GCE (CV c) has its peak current sharply increased. The ($\Delta E_p = E_{pa} - E_{pc}$) calculated for CT and HQ at

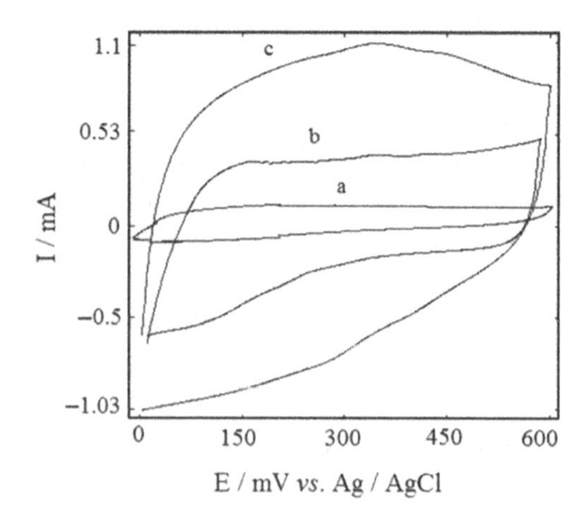

Figure 3.26 CVs of the (a) bare GCE, (b) GR–GCE, and (c) P4VP/GR–GCE in 0.1 M sodium sulfate buffer (pH 2.5) at scan rate 20 mV s^{-1}.

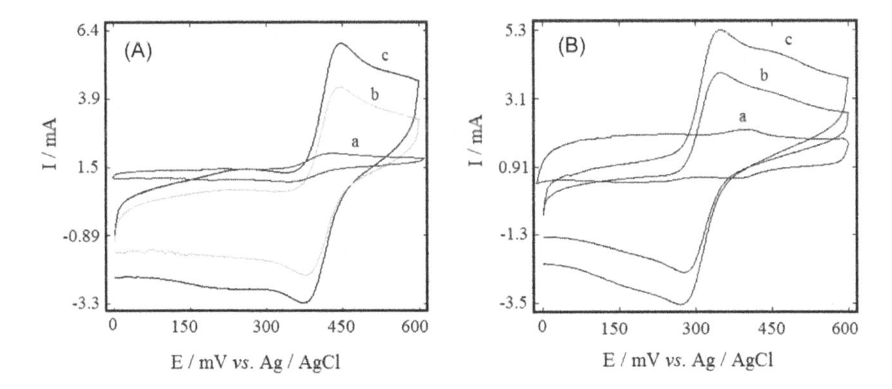

Figure 3.27 CVs of (A) 2 μM CT and (B) 2 μM HQ at the (a) bare GCE, (b) GR–GCE, and (c) P4VP/GR–GCE in 0.1 M sodium sulfate buffer (pH 2.5) at scan rate 20 mV s^{-1}.

the P4VP/GR–GCE are about 70 and 100 mV, respectively. While at GR–GCE, ΔE_p values for CT and HQ are 90 and 130 mV, respectively.

Hence, smaller ΔE_p for CT and HQ at the P4VP/GR–GCE indicates the electrode process is highly reversible as a result of increase in kinetics of electron transfer [139] as and when P4VP is present in the nanocomposite-modified electrode as compared to the GR–GCE and bare GCE. Moreover, based on EIS study (see Section 3.8.3), the R_{ct} indicates that the kinetic of charge transfer at the P4VP/GR–GCE is very favorable. These are the evidences of the electrocatalytic effects of the P4VP/GR–GCE toward oxidation of CT and HQ. The synergistic effect of GR and P4VP in P4VP/GR nanocomposite has provided an effective microenvironment for the redox process of these diphenols. As a result, it is then concluded that incorporation of P4VP with electron mediator property into composite along with remarkable physical properties of GR can be effective in increasing the sensitivity of nanocomposite-modified electrode toward CT and HQ.

The E_{pa}, of each CT (460 mV) and HQ (340 mV), clearly indicates anodic peak separation, ΔE_p, value of 120 mV which is favorable for simultaneous determination of the diphenol isomers. The CV of the mixture of HQ and CT is shown in Fig. 3.28. At bare GCE (CV a), the electrochemical response only shows baseline, while at the GR–GCE (CV b) despite the increase in anodic response, it is still not possible to separate and detect the HQ and CT in the mixture. However, at P4VP/GR–GCE (CV c), two well-defined anodic peaks

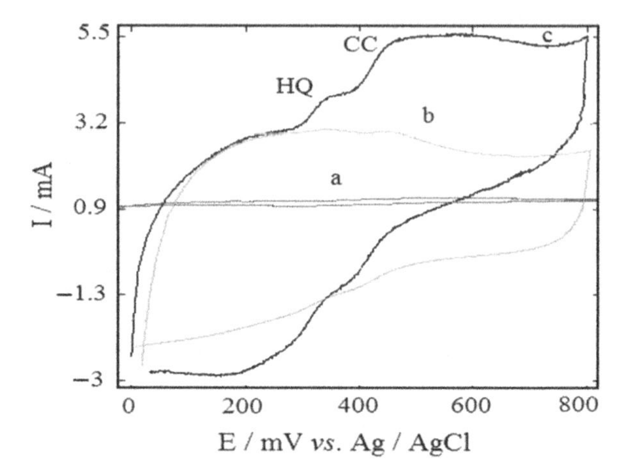

Figure 3.28 CVs of 2 μM HQ and 2 μM CT in 0.1 M sodium sulfate buffer (pH 2.5) at (a) bare GCE, (b) GR–GCE, and (c) P4VP/GR–GCE at scan rate 20 mV s^{-1}.

correspond to HQ and CT appears at 340 and 460 mV, respectively. The I_{pa} of HQ and CT on the P4VP/GR–GCE are significantly higher than that on the bare GCE and GR–GCE. Hence, the P4VP/GR–GCE facilitates electron transfer, catalyzed the electro oxidation of the diphenols studied. Thus, similar mechanism to Scheme 3.2, for oxidation of the diphenols to quinones in the $E_a = 460$ for CT and $E_a = 340$ for HQ at P4VP/GR–GCE, is suggested (see Section 3.9) [235]. The suggested scheme shows that P4VP/GR–GCE can act as an electron transfer mediator to accelerate the electron transfer rate for the oxidation of CT and HQ. In addition, the composite film with different mass ratio of GR and P4VP (i.e., 4:2, 2:4, 1:1, 3:2, and 2:3) were fabricated (Fig. 3.29). Then, the effects of these ratios on voltammetric responses of the nanocomposite electrodes for the detection of HQ and CT were investigated. As can be seen, a well-defined redox peak with a significant I_{pa} is obtained in the mass ratio of 4:2 of GR and P4VP. In this regard, the effect of pH and scan rate has been studied on the modified electrode at optimum mass ratio of GR and P4VP (4:2).

3.14.1 Effects of Solution pH

The effect of solution pH on the electrochemical behavior was investigated using CV. Variations of peak currents with respect to pH of the electrolyte in the pH range from 1.5 to 7 for CT and HQ is shown

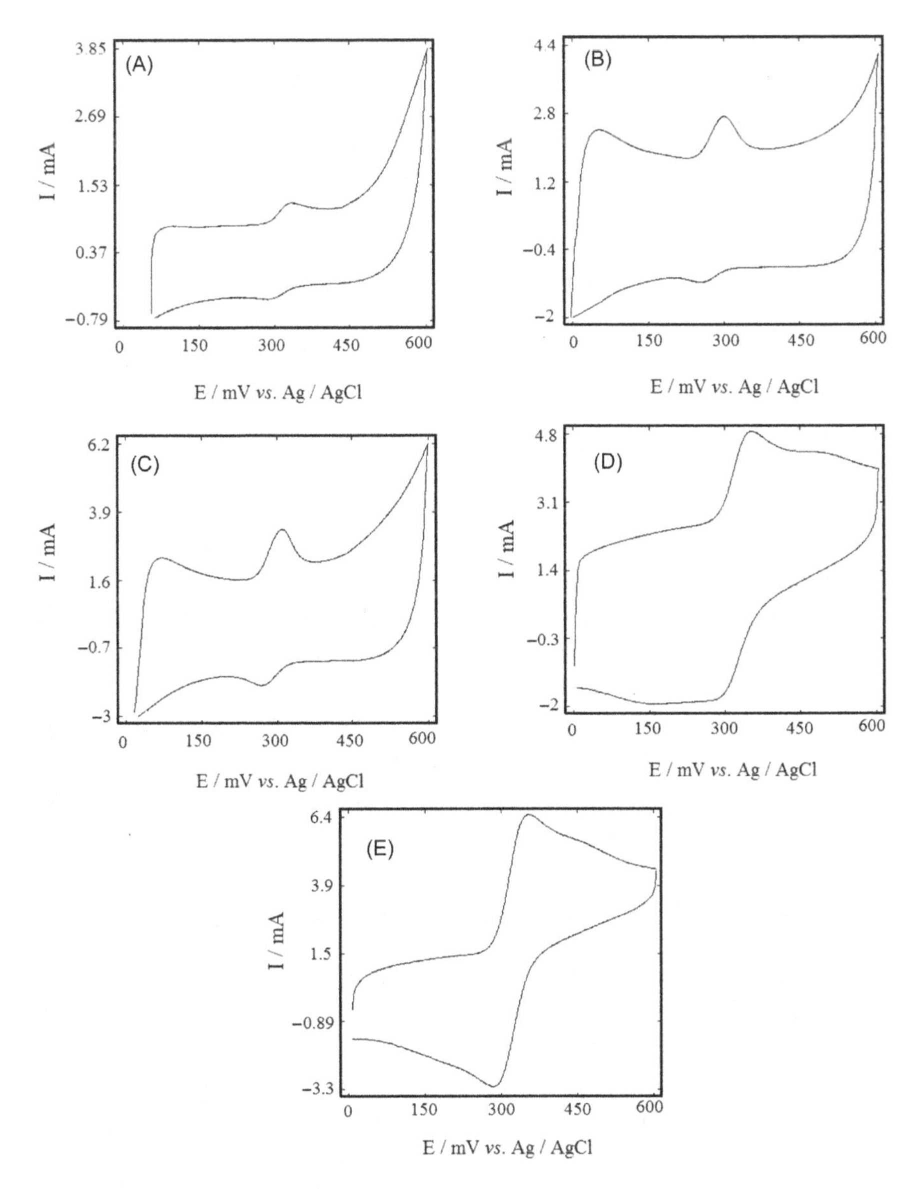

Figure 3.29 CVs of (A) 2 μM CT at different mass ratio of GR and P4VP (A) 2:4, (B) 3:2, (C) 1:1, (D) 2:3, and (E) 4:2 in 0.1 M sodium sulfate buffer (pH 2.5) at scan rate 20 mV s^{-1}.

(Fig. 3.30). The results revealed that I_{pa} of CT and HQ increases with solution pH until the pH reaches 2.5. However, less current was detected when the pH solution was lower or higher than 2.5 which suggest a kinetically less favorable reaction at lower and higher pH than 2.5. The decrease in peak current after pH 2.5 could be related to a

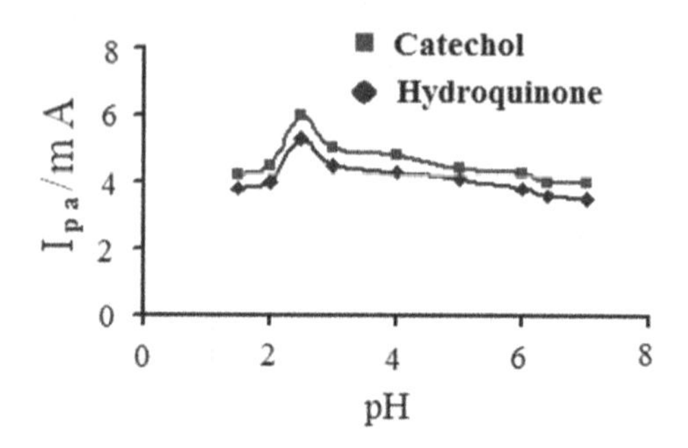

Figure 3.30 Effect of the pH on the anodic peak currents.

decline in the rate of the coupling reaction between CT and *o*-quinone [290]. The results illustrate that highest electrochemical response for the oxidation of the diphenols is at pH 2.5.

3.14.2 Effect of Scan Rate

The scan rate dependence of the modified electrode for the diphenols was investigated (Fig. 3.31). As the scan rate increases, the I_{pa} is increased and the E_{pa} is shifted positively, especially, at scan rate higher than 100 mV s^{-1}. This could be due to changes in the electrocatalytic activity and kinetic effect of GR/GCE surface on the oxidation of the diphenols. The I_{pa} has a linear relationship with the $v^{1/2}$ indicating that the oxidations of HQ and CT are diffusion-controlled processes.

3.15 DETERMINATION OF CT AND HQ USING DPV

Since DPV has better current sensitivity and resolution than CV, it is then utilized for quantitative determination of CT and HQ (calibration plots) at P4VP/GR–GCE. In the individual DPV (Fig. 3.32) of various concentrations of CT (Fig. 3.32A) and HQ (Fig. 3.32B) in 0.1 M sodium sulfate buffer (pH 2.5), the I_{pa} has a linear relationship with the concentration of CT and HQ in the range of 0.1 to 10 μM. Two linear equations I_{pa} (mA) = 0.58 [CT] (μM) + 3.7 and I_{pa} (mA) = 0.66 [HQ] (μM) + 3.3 with linear regression coefficients R^2 of 0.990 and 0.985, respectively, are obtained. The detection limits of HQ and CT are 8.1 and 26 nM, respectively. However, Fig. 3.32C shows the experiment on the P4VP/GR–GCE for the simultaneous determination of HQ and CT by continuously changing their concentrations. The DPV results indicate well-separated peaks at potentials 340 and 460 mV which correspond to the HQ and CT, respectively. This confirms that

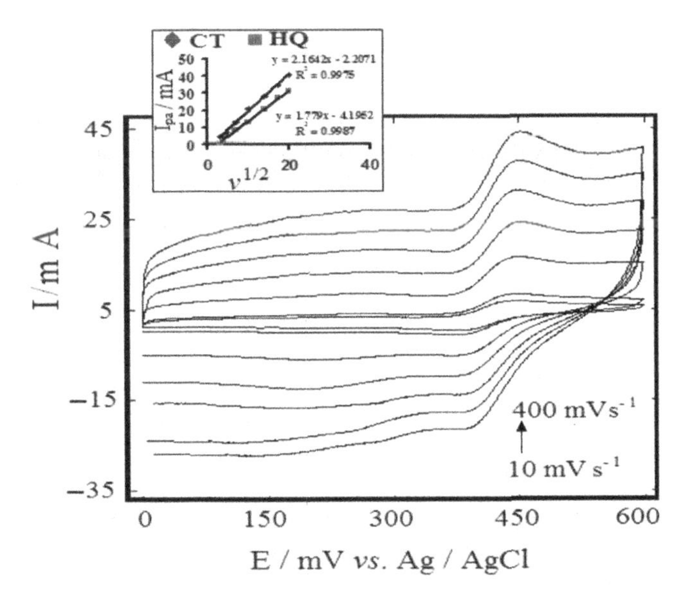

Figure 3.31 CVs of 2 μM CT in sodium sulfate buffer (pH 2.5) at P4VP/GR−GCE at scan rates of 10, 20, 50, 100, 200, 300, 400 mV s⁻¹. Inset: Linear relationship of anodic peak current of 2 μM CT and HQ versus square roots of scan rates.

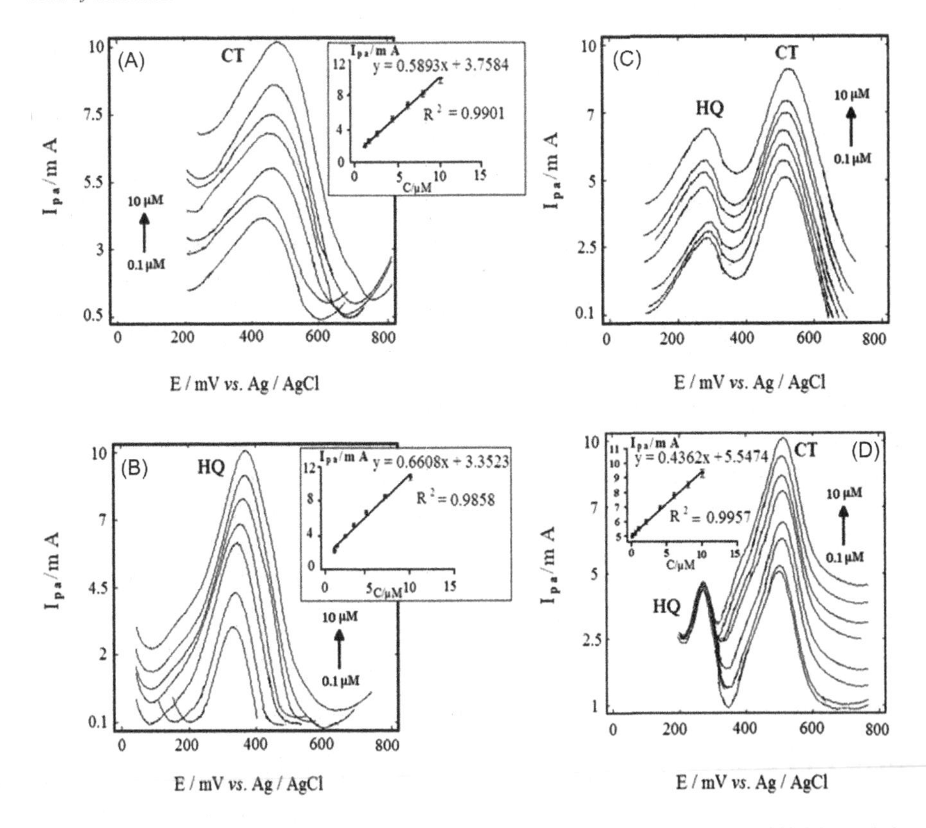

Figure 3.32 DPVs of (A) 0.1, 0.4, 1, 2, 6, 8, 10 μM CT; (B) 0.1, 0.4, 2, 3, 5, 7, 10 μM HQ; (C) 0.1, 0.5, 1, 2, 4, 8, 10 μM HQ and CT; and (D) 0.1, 0.4, 1, 2, 4, 6, 8, 10 μM CT in the presence of 2 μM HQ at P4VP/GR−GCE in sodium sulfate buffer (pH 2.5).

Table 3.9 Responses of Some of HQ and CT Sensors Constructed from Various Electrodes Material

Electrode Materials	Technique	Detection Limit	Linear Range (μM)	Ref.
Graphene	DPV	0.8 μM HQ	1–10 and 10–80	[229]
ECF–CPE	DPV	0.4 μM HQ	1–200	[200]
		0.2 μM CT		
Poly(glutamic acid)/GCE	DPV	1 μM HQ	5–80	[203]
		0.8 μM CT	1–80	
p-Phenylalanine-modified electrode	DPV	1.0 μM HQ	10–140	[236]
		0.7 μM CT		
Electrochemically activated GCE	SWVs	0.018 μM HQ	1–100	[238]
		0.032 μM CT	2–100	
Graphene/BMIMPF$_6$ nanocomposite-modified electrode	DPV	0.01 μM HQ	0.5–50	[69]
		0.02 μM CT		
P4VP/GR–GCE	DPV	8.1 nM HQ	0.1 to 10	Present study
		26 nM CT		

the developed electrode is suitable for the simultaneous detection of CT and HQ in mixed system of diphenols. Table 3.9 shows the comparison between the results for the determination of dihydroxybenzene isomers. The limit of detections of HQ and CT at P4VP/GR–GCE revealed the electrode's better sensitivity than other electrodes.

DPV was also performed at the P4VP/GR–GCE for the individual determination of HQ and CT in the mixtures when the concentration of one species changes and the other one is fixed (Fig. 3.32D). Typical DPVs show different concentrations of CT in 0.1 M sodium sulfate buffer (pH 2.5) with 2 μM HQ. The I_{pa} of CT at 460 mV increases linearly as the concentration of CT changes from 0.1 to 10 μM while the I_{pa} of HQ is unchanged. Similar behavior has been reported [234,235] for HQ with respect to CT without any mutual interference.

3.16 APPLICATION TO REAL SAMPLE ANALYSIS

The developed electrode was tested for the simultaneous determination of CT and HQ in synthetic water samples (tap water and lake water).

Table 3.10 Recoveries of CT and HQ in Tap Water ($n = 5$)

Sample	Tap Water Containing CT	HQ Added (μM)	CT Found (μM)	Recovery (%)	RSD (%)
1	4.30	0	4.23	98.3	3.5
2	4.30	10	14.13	98.8	1.4
3	4.30	20	24.53	100.9	1.6
Sample	**Tap Water Containing HQ**	**CT Added (μM)**	**HQ Found (μM)**	**Recovery (%)**	**RSD (%)**
1	2.22	0	2.23	100.4	2.5
2	2.22	10	11.96	97.8	1.7
3	2.22	20	21.80	98.1	1.9

Table 3.11 Recoveries of CT and HQ in Lake Water ($n = 5$)

Sample	Lake Water Containing CT	HQ Added (μM)	CT Found (μM)	Recovery (%)	RSD (%)
1	3.53	0	3.46	98. 0	2.6
2	3.53	10	13.50	99.7	2.6
3	3.53	20	23.86	101.4	3.2
Sample	**Lake Water Containing HQ**	**CT Added (μM)**	**HQ Found (μM)**	**Recovery (%)**	**RSD (%)**
1	1.65	0	1.68	101.8	2.0
2	1.65	10	11.32	97.1	2.0
3	1.65	20	21.49	99.2	2.0

Since the amount of HQ and CT are unknown in water samples, the standard addition method is applied by spiking with known concentrations of the analytes into the samples. The recovery was obtained by measuring DPV responses for samples in which CT and HQ are added. The RSD of this method, based on five replicates ($n = 5$), is presented in Tables 3.10 and 3.11. Satisfactory recoveries of CT and HQ at P4VP/GR–GCE in the range of 0.1 to 10 μM reveal that this method is effective and reliable. These findings indicate that the method is suitable for the effective and sensitive analysis of CT and HQ.

3.17 INTERFERENCE STUDIES

Phenol, nitrophenol, aminophenols, bisphenol A, and chlorophenols are the common interferences in the determination of 2 μM CT and HQ. The CVs of these compounds at different concentrations from 0.1 to 10 μM on the surface of modified electrode in the potential range of

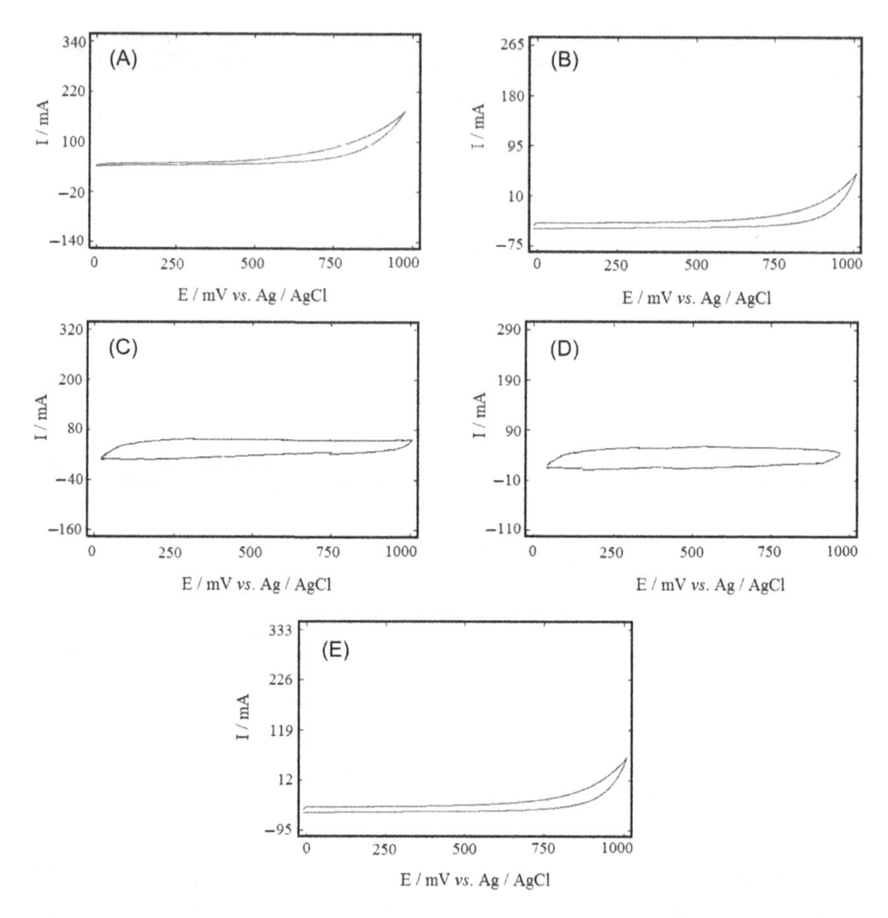

Figure 3.33 CVs of (A) nitrophenol, (B) phenol, (C) aminophenols (o-, m-, p-), (D) bisphenol A, and (E) chlorophenols at different concentrations (0.1 to 10 μM) on the modified electrode at potential range of 0 to 1.0 V.

0 to 1 V show that these compounds do not interfere even in large amount (Fig. 3.33). Consequently, the coexistence of these phenolic compounds do not interfere with the quantitative detection of HQ and CT. In addition, no significant interference from common cations such as Ca^{2+}, Mg^{2+}, Mn^{2+}, Zn^{2+}, Fe^{3+}, Cu^{2+}, and Co^{2+} was observed.

3.18 REPRODUCIBILITY AND STABILITY OF P4VP/GR–GCE

The regeneration of the surface of electrode was examined by cyclic voltammetric studies of seven different electrodes constructed by the same procedure. A RSD of 3.8% ($n = 7$) is obtained indicating good reproducibility. In addition to this, the long-term stability of the

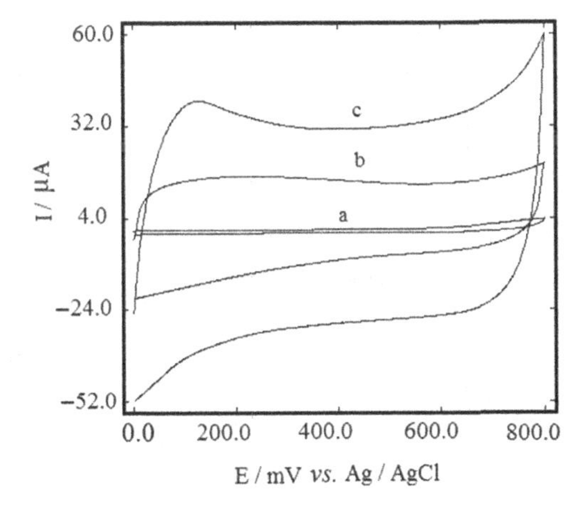

Figure 3.34 CVs of (a) bare GCE, (b) MWCNT−GCE, and (c) P4VP/MWCNT−GCE in 0.1 M phosphate buffer (pH 7) at scan rate 20 mV s^{-1}.

modified electrode was also evaluated. The modified electrode was constantly used for 2 months. It has retained 99% of its current response to the diphenols studied.

3.19 ANALYSIS OF PHARMACEUTICAL SAMPLE

The proposed electrode (P4VP/MWCNT−GCE) was tested in pharmaceutical samples, to develop a sensitive and accurate method to determine analgesic drugs. The following discussion demonstrates the analytical performance of this electrode in sensing of some of these drugs such as paracetamol (PCT), aspirin (ASA), and caffeine (CF).

3.20 ELECTROCHEMICAL BEHAVIOR OF PCT

At bare GCE (Fig. 3.34a) and in the absence of PCT, there is poor electrochemical response showing only a capacitive current. However, this peak current is more noticeable for the MWNT−GCE (Fig. 3.34b) and particularly for P4VP/MWCNT−GCE (Fig. 3.34c) indicating that the effective surface area of the P4VP/MWCNT−GCE is larger than that of MWNT−GCE.

The electrochemical response of 100 µM PCT in 0.1 M phosphate buffer (pH 7) at three different electrodes have been studied by using CV (Fig. 3.35). The broad redox couple peaks of PCT at the bare GCE and MWNT−GCE (Fig. 3.35a and b) could indicate a sluggish

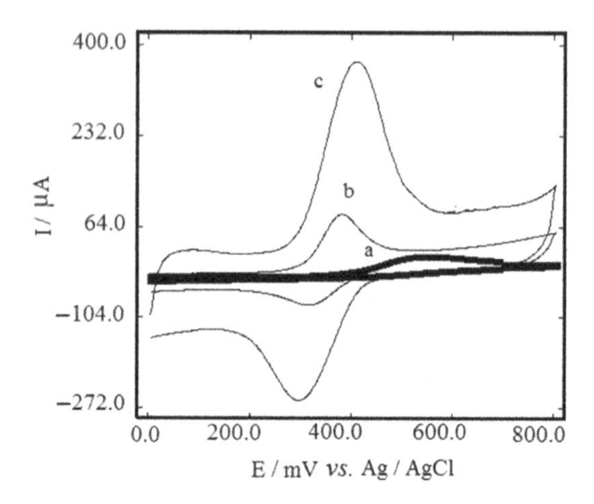

Figure 3.35 CVs of 100 μM PCT at the (a) bare GCE, (b) MWCNT−GCE, and (c) P4VP/MWCNT−GCE in 0.1 M phosphate buffer (pH 7). Scan rate 20 mV s^{-1}.

rate of electron transfer. However, the P4VP/MWCNT−GCE displays well-defined redox peaks with E_{pa} at 404 mV and E_{pc} at 307 mV. The ΔE_p (97 mV) indicates a favorable quasireversible electrode process (Fig. 3.35c). It is obvious that the increase in peak currents is due to the huge increment in the area of the electrode surface modified with P4VP/MWCNT.

In addition, the composite films with different mass ratio of MWCNT and P4VP at 2:4, 2:3, 1:1, 3:2, and 4:2 were fabricated. Then, the effects of these ratios on voltammetric responses of the nanocomposite electrodes for the detection of PCT were investigated (Fig. 3.36) (see Section 3.9).

3.20.1 Effects of Solution pH

The effect of solution pH on the electrochemical response of the P4VP/MWCNT−GCE toward 100 μM PCT was investigated using CV (Fig. 3.37). Variation of peak currents with respect to pH of the electrolyte in the pH range of 2.5 to 8 is shown in Fig. 3.38A. It can be seen that the I_{pa} increases with solution pH until the pH reaches 7. But in Fig. 3.35, the I_{pa} of PCT in phosphate buffer (pH 7) is higher than the reduction peak current, I_{pc}, because PCT is quasireversible process. Consequently, the buffering at pH 7, which is near to the physiological pH, will be used for the rest of the work. A small current was detected when the pH of the solution was either lower or higher than 7. The broadening of oxidation peak potential, and decreasing of I_{pa} are

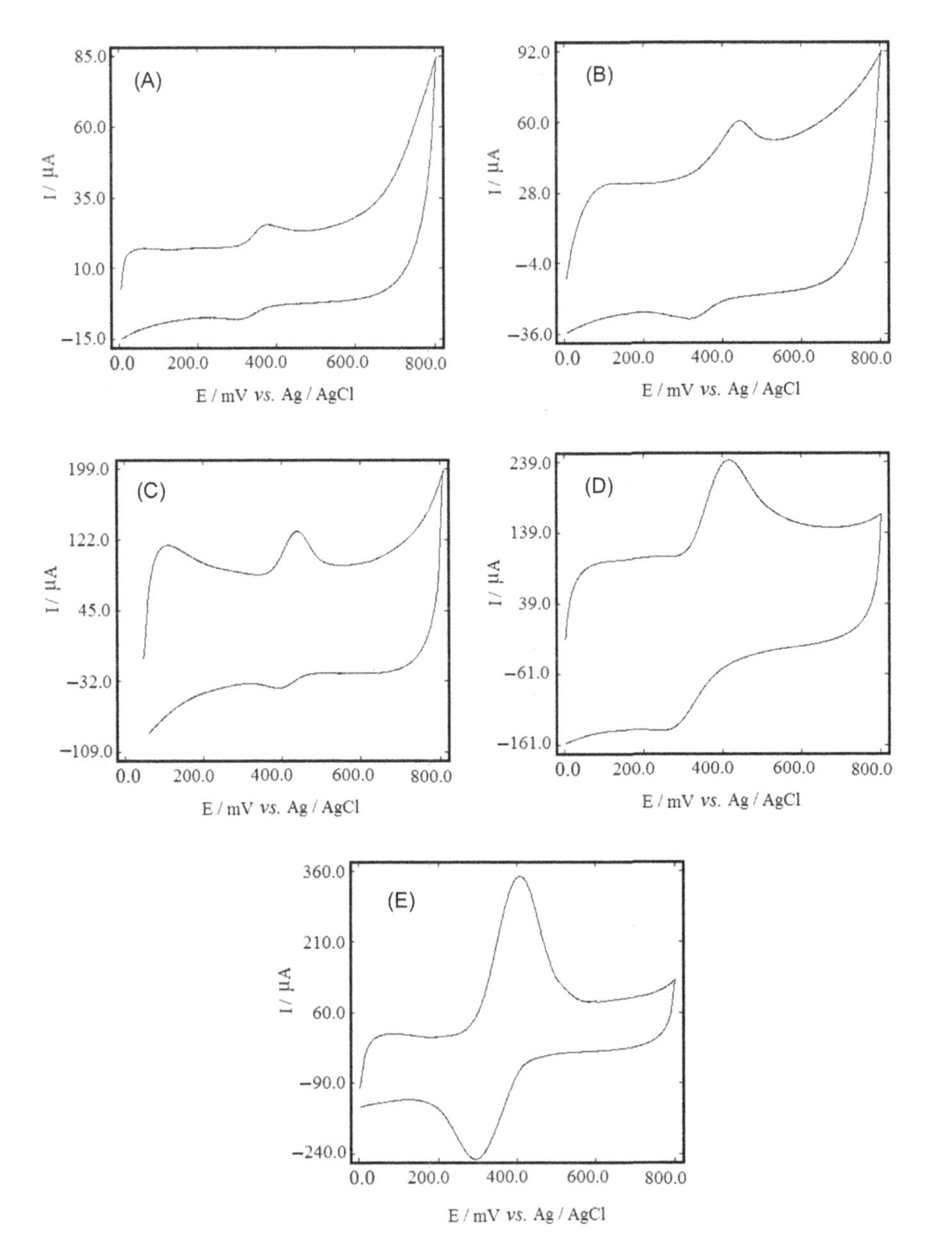

Figure 3.36 CVs of (A) 100 μM PCT at the different mass ratio of MWCNT and P4VP (A) 2:4, (B) 2:3, (C) 1:1, (D) 3:2, and (E) 4:2 in phosphate buffer (pH 7) at P4VP/MWCNT−GCE. Scan rate 20 mV s^{-1}.

observed with increasing basicity, at pH higher than 7, suggesting a kinetically less favorable reaction at higher pH. Besides, the protonated form of PCT is more soluble in aqueous solution than the neutral species. The effect of phosphate buffer pH on E_{pa} has been investigated

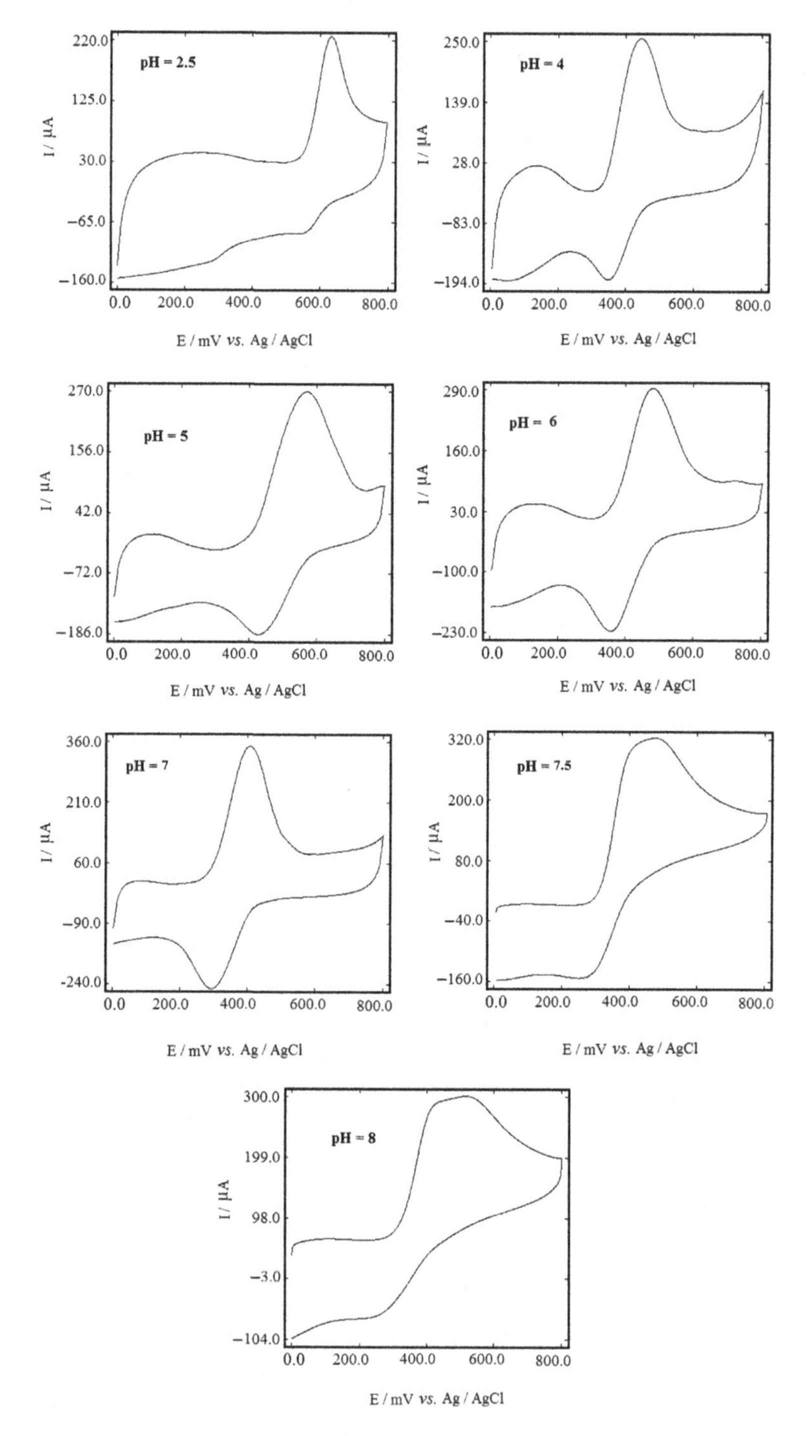

Figure 3.37 CVs of 100 μM PCT at P4VP|MWCNT−GCE at various pH. Scan rate 20 mV s^{-1}.

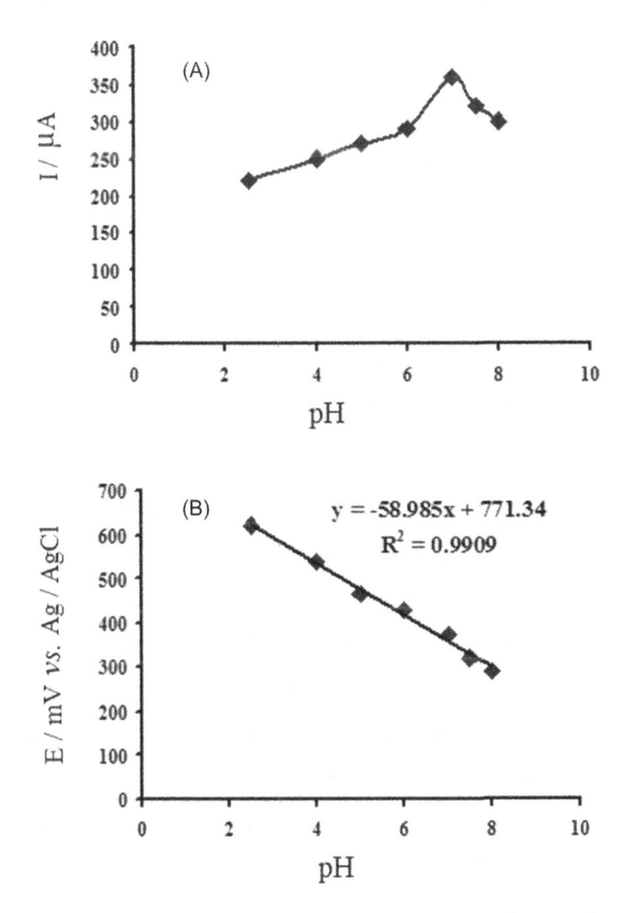

Figure 3.38 Effect of the pH on the (A) anodic peak currents and (B) anodic peak potential.

in the mentioned range. The results show that the E_{pa} is shifted toward negative potentials with a slope of -58 mV decade^{-1}. A linear relationship of E_{pa} (V) $= -0.058$ pH $+ 771.34$ is obtained with ($R^2 = 0.990$) (Fig. 3.38B). The slope is very close to the Nernstian value of -59 mV decade^{-1}. This suggests that the number of protons and electrons transferred in the redox reaction of PCT are equal and likely to be two [292,293]. Thus, the mechanism in Scheme 3.3 for redox of PCT is proposed [292–296].

3.20.2 Effect of Scan Rate

The scan rate dependence of the modified electrode for 100 μM PCT was also investigated (Fig. 3.39). It is clear that the redox peak currents

Scheme 3.3 The expected redox mechanism for PCT.

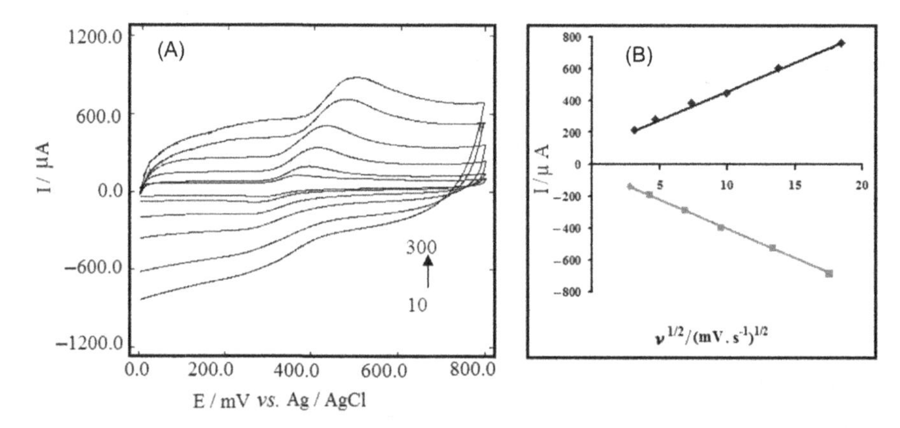

Figure 3.39 (A) CVs of 100 μM PCT in 0.1 M phosphate buffer (pH 7) at P4VP/MWCNT−GCE at scan rates 10, 20, 50, 100, 200, 300 mV s^{-1} and (B) Linear relationship of anodic and cathodic peaks currents for 100 μM PCT versus square root of scan rate.

at the P4VP/MWCNT−GCE in the PCT solution increase linearly as the scan rate increases from 10 to 300 mV s^{-1}. In addition, E_{pa} is shifted slightly to positive potentials, while E_{pc} is negatively shifted. This could be due to changes in the electrocatalytic activity and kinetic effect of P4VP/MWCNT−GCE surface on the oxidation of PCT especially at scan rates higher than 100 mV s^{-1}. In other words, at scan rates higher than 100 mV s^{-1}, the time window for the PCT oxidation becomes very narrow, avoiding the facile electron transfer between substrate and catalytic sites. The linear relationship between the peak current and scan rate can be expressed by the linear regression equation as: $I_{pa}/\mu A = 26.10X + 211.26\ \nu/\text{mV s}^{-1}$ ($R^2 = 0.994$) and $I_{pc}/\mu A = -32.76X - 91.86\ \nu/\text{mV s}^{-1}$ ($R^2 = 0.996$), respectively. The results indicate that the electrochemical reaction of PCT on the P4VP/MWCNT−GCE is a surface-controlled process [297,298].

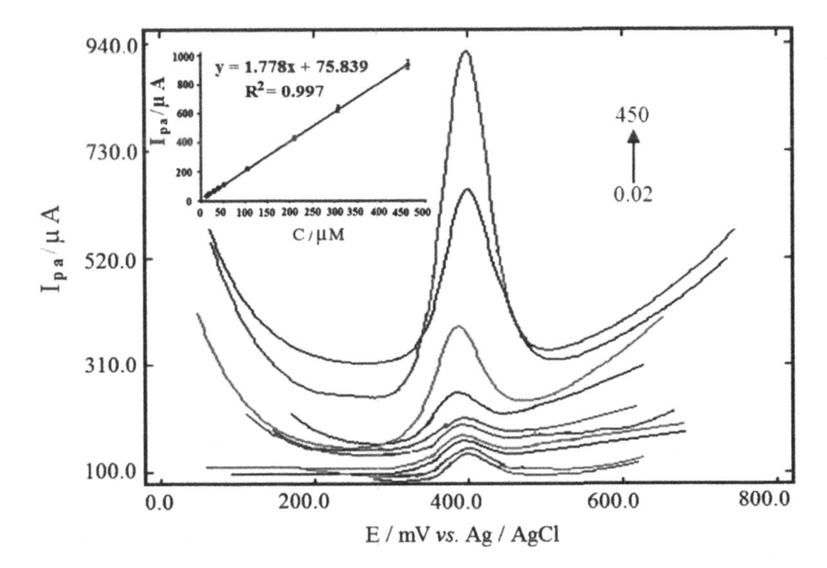

Figure 3.40 DPVs of 0.02, 0.04, 2, 10, 20, 50, 100, 200, 300, 450 µM PCT in 0.1 M phosphate buffer (pH 7) at P4VP/MWCNT–GCE. Inset: The corresponding calibration plot.

3.21 DETERMINATION OF PCT BY DPV

DPV is superior to CV because of its sensitivity, resolution and lower limit of detection. Fig. 3.40 shows the DPV of different concentrations of PCT in 0.1 M phosphate buffer (pH 7) with applied potentials of 0 to 0.8 V, step potential of 2.0 mV, modulation amplitude of 50 mV and a scan rate of 10 mV s^{-1}. The I_{pa} is linearly proportional to the concentration of PCT in the range of 0.02 to 450 µM. A linear equation of I_{pa} (µA) = 1.778[PCT] (µM) + 75.839, with $R^2 = 0.997$, was obtained. The detection limit of PCT at P4VP/MWCNT–GCE is 1.7 nM which is superior to previous reports (Table 3.12).

3.22 ANALYSIS OF REAL SAMPLES

3.22.1 Determination of PCT in Formulation Tablets

The developed electrode was tested for the determination of PCT in tablets. The recovery was obtained by using DPV to evaluate the accuracy of the method. The RSD of this method, based on five replicates ($n = 5$), is presented in Table 3.13. Recoveries of PCT at P4VP/MWCNT–GCE in the range of 0.02 to 450 µM reveal that this method is effective and reliable. These findings indicate that the method is rapid and simple for the selective and sensitive analysis of PCT in pharmaceutical preparations.

Table 3.12 Responses of Some of PCT Sensors Constructed from Various Materials

Electrode Materials	Technique	Detection Limit	Linear Range (μM)	Ref.
Carbon ionic liquid electrode	DPV	0.3 μM	1–2000	[299]
Gold-nanoparticle-modified CPE	DPV	0.0146 μM	0.05–270	[300]
Poly(thaurine)/MWCNT-modified GCE	DPV	0.5 μM	1–100	[242]
SWNT/DCP-modified GCE	SWV	0.04 μM	0.1–20	[240]
SPE/PEDOT	DPV	1.39 μM	4–400	[253]
MWCNTs	DPV	39.8 nM	0.074–230	[251]
MWCNT–BPPGE	CV	45 nM	0.1–25	[301]
MWCNT–BPPGE	SWASV	10 nM	0.01–2	[301]
MWCNT/Li^+/Bi_2O_3 Composite-modified GCE	CV	0.74 μM	0.5–2000	[294]
Poly(DA)/GC	CV	0.0067 μM	0.02–500	[302]
MWCNT-L-Cys-Au/SAMs-CME	FIIB	1 μM	2–1000	[303]
ETPG	ATSDPV	0.0025 μM	0.05–2.5	[245]
MWCNT/GCE	CV	0.02 μM	0.1–1000	[304]
Carbon paste electrode modified with CNT/ poly (3-aminophenol)	SWV	1.1 μM	10–100	[305]
GCE/MWCNT–polyhistidine	DPV	32 nM	0.25–10	[297]
P4VP/MWCNT–GCE	DPV	1.7 nM	0.02–450	Present study

FIIB = flow injection irreversible biamperometric.

Table 3.13 Determination of PCT in Formulation Tablets Using P4VP/MWCNT–GCE ($n = 5$)

Sample	PCT (μM)			RSD (%)	Recovery (%)
	Contents	Spike	Found		
1	81.45	0	80.90	1.4	99.3
2	81.45	10	90.03	1.5	98.4
3	81.45	20	100.66	1.5	99.2

3.22.2 Determination of PCT in Urine Samples

P4VP/MWCNT–GCE was investigated for the measurement of PCT in three human urine samples. The percentage of recovery of the spiked sample is in the range between 99.1 and 101.3 (Table 3.14). This shows that the modified electrode is suitable for the determination of PCT in biological fluids.

Table 3.14 Determination of PCT in Human Urine Samples Using P4VP/MWCNT–GCE ($n = 3$)

Sample	PCT (μM)			RSD (%)	Recovery (%)
	Contents	Added	Found		
1	94.44	25	121.06	1.1	101.3
2	96.70	25	121.66	1.0	99.9
3	98.40	25	122.33	1.2	99.1

3.23 REPRODUCIBILITY AND STABILITY OF P4VP/MWCNT–GCE

The electrode reproducibility was examined using CV over seven electrodes constructed by the same procedure. A RSD of the I_{pa} is 3.6% which indicates good reproducibility. The operational and storage stabilities of the P4VP/MWCNT–GCE for oxidation of 100 μM PCT in both synthetic and real samples have also been studied. Long term stability is obtained when the modified electrode is kept in 0.1 M phosphate buffer (pH 7) at 4°C when not in use. The operational stability is retained at 99% of the initial current when it is subjected under optimum conditions to approximately 400 cycles and more than 60 days of continuous use.

3.24 INTERFERENCE STUDIES

A sensor which is to be used in biological samples must be able to discriminate between target analytes and interfering species such as AA and UA commonly present in physiological environments. It is difficult to detect PCT in the presence of AA and UA by electrochemical methods because they have nearly the same redox potential ranges and comparable sensitivities on the unmodified electrode [306]. The effects of AA and UA on the measurement of PCT were investigated using CV. Fig. 3.41a and a indicates that E_{pa} of a mixture of AA, UA, and PCT is inseparable at the bare GCE and MWCNT–GCE. At the P4VP/MWCNT–GCE, the E_{pa} of AA is close to that of UA at the same potential range. The well-defined wave of PCT was obtained at the P4VP/MWCNT–GCE with good separation from AA and UA. The anodic peak potentials of AA, UA and PCT were 290, 300 and 404 mV, respectively. This indicates that the voltammetric determination of PCT in biological samples is devoid of any interference from AA and UA.

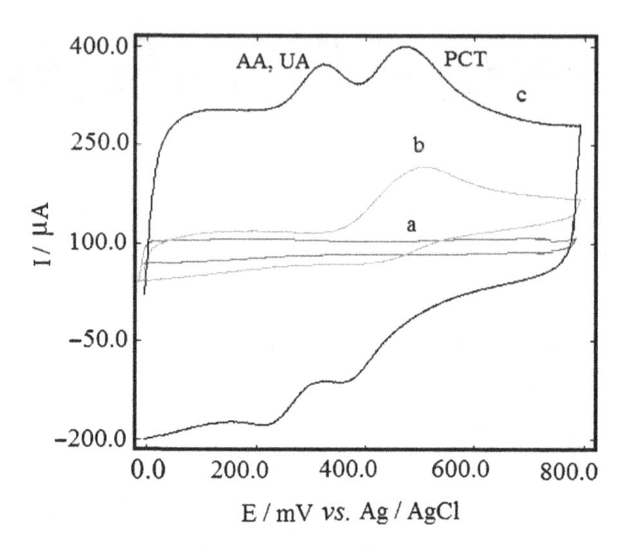

Figure 3.41 CVs of 100 μM AA, 100 μM UA and 100 μM PCT at the (a) bare GCE, (b) MWCNT–GCE, and (c) P4VP/MWCNT–GCE in 0.1 M phosphate buffer (pH 7) at scan rate of 20 mV s^{-1}.

3.25 ELECTROCHEMICAL BEHAVIOR OF PCT ON THE P4VP/GR–GCE

Fig. 3.42 shows CVs obtained for the bare GCE, GR–GCE, and P4VP/GR–GCE electrodes in supporting electrolyte (buffer pH 7). A bare GCE in the absence of PCT has only capacitive current and with poor electrochemical response (Fig. 3.42a). However, for the GR–GCE (Fig. 3.42b) and particularly for P4VP/GR–GCE (Fig. 3.42c) the peak current is more noticeable. This indicates that greater electroactive surface area of the P4VP/GR–GCE.

The electrochemical response of 100 μM PCT in 0.1 M phosphate buffer (pH 7) at three different electrodes have been studied by using CV (Fig. 3.43). The broad redox couple peaks of PCT at the bare GCE and GR–GCE (Fig. 3.43a and b) could indicate a sluggish rate of electron transfer. However, the P4VP/GR–GCE displays well-defined redox peaks with E_{pa} at 441 mV and E_{pc} at 369 mV. The ΔE_p calculated for PCT at the P4VP/GR–GCE is 72 mV indicates a quasireversible electrode process (Fig. 3.43c). While at GR–GCE, ΔE_p values for PCT is 111 mV. Hence, smaller ΔE_p for PCT at the P4VP/GR–GCE indicate the higher reversibility of the electrode process as a result of faster kinetics of electron transfer [138] when P4VP is present in the nanocomposite-modified electrode as compared to the GR–GCE and

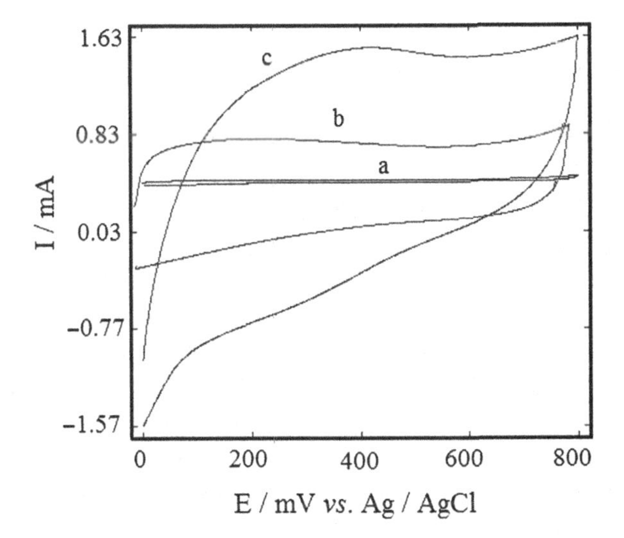

Figure 3.42 CVs obtained for the (a) bare GCE, (b) GR−GCE, and (c) P4VP/GR−GCE in 0.1 M phosphate buffer solution (pH 7) at scan rate of 20 mV s^{-1}.

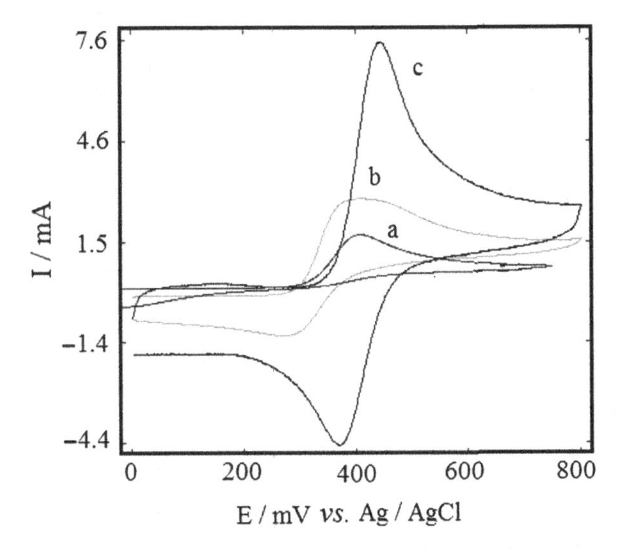

Figure 3.43 CVs of 100 μM PCT at the (a) bare GCE, (b) GR−GCE, and (c) P4VP/GR−GCE in 0.1 M phosphate buffer solution (pH 7) at scan rate 20 mV s^{-1}.

bare GCE. Moreover, based on EIS studies the R_{ct} indicates that the kinetics of charge transfer at the P4VP/GR−GCE is very favorable. It is obvious that the increase in peak currents is due to the huge increment in the area of the electrode surface modified with P4VP and GR.

3.25.1 Effect of Solution pH

As a proton takes part in the electrode reaction process of PCT, the effect of solution pH on the electrochemical behavior of 100 μM PCT at the P4VP/GR–GCE was investigated using CV (Fig. 3.44). Variation of peak currents with respect to pH of the electrolyte in the pH range of 2.5 to 8 is shown in Fig. 3.45A. It can be seen that the I_{pa} increases with solution pH until the pH reaches 7. Moreover, the buffering at pH 7, which is near to the physiological pH, will be used for the rest of the work. A small current was detected when the pH of the solution was either lower or higher than 7. The decrease of I_{pa} is observed with increasing basicity, at pH higher than 7, suggesting a kinetically less favorable reaction at higher pH. Similar to Section 3.20.1, the effect of phosphate buffer pH on E_{pa} has been investigated in the mentioned range. The results show that the E_{pa} is shifted toward negative potentials with a slope of -57 mV decade^{-1}. A linear relationship of E_{pa} (V) = -0.057 pH + 715.04 is obtained with ($R^2 = 0.994$) (Fig. 3.45B). The slope is very close to the Nernstian value of -59 mV decade^{-1}. This suggests that the number of protons and electrons transferred in the redox reaction of PCT are equal and likely to be two [252,292]. Thus, similar mechanism to Scheme 3.3 for oxidation of PCT in $E_{pa} = 441$ mV is proposed [252,293,296].

3.25.2 Influence of Scan Rate

Fig. 3.46A shows the CV of 100 μM PCT at the P4VP/GR–GCE at different scan rates. With the increase of scan rate from 10 to 400 mV s^{-1}, the I_{pa} increased gradually along with the ΔE_p values. Then a linear relationship between the peak current (I_{pa}) and the scan rate ($v^{1/2}$) was plotted (Fig. 3.46B) with the regression equation as: I_{pa}/mA = 3.33X − 4.52 ν/mV s^{-1} ($R^2 = 0.998$) and I_{pc}/mA = − 1.57X − 4.22 ν/mV s^{-1} ($R^2 = 0.986$), respectively. It indicates that a diffusion-controlled process occurs at the electrode. Hence, it is obvious that the P4VP/GR–GCE possess faster charge-transfer kinetics which is attributed to the presence of P4VP and GR as modifiers. The results indicate that the electrochemical reaction of PCT on the P4VP/GR–GCE is a surface-controlled process [297,298].

3.26 DETERMINATION OF PCT BY DPV

DPV was applied for the identification of the PCT in 0.1 M phosphate buffer (pH 7) with applied potentials of 0 to 0.8 V, step potential of

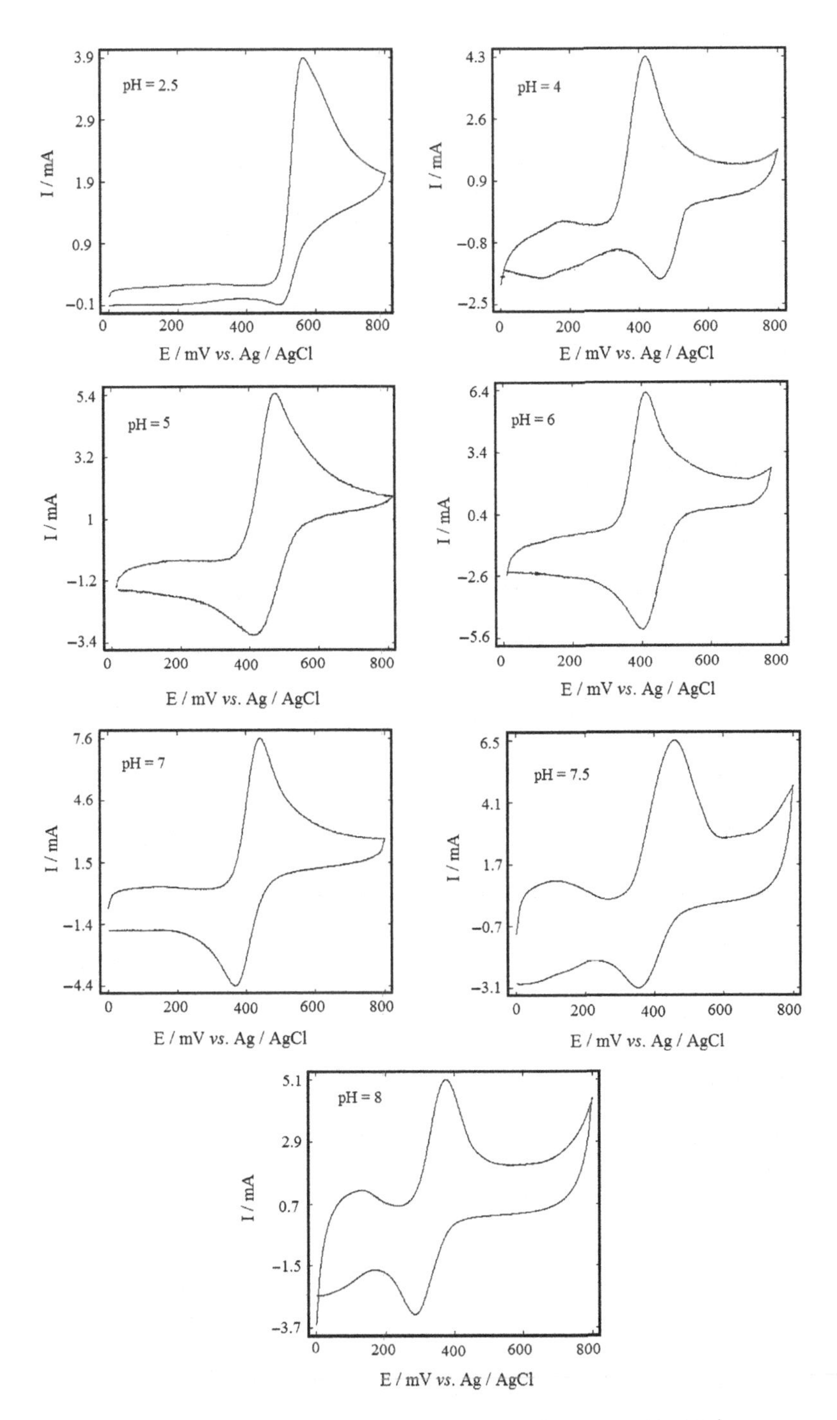

Figure 3.44 CVs of 100 μM PCT at P4VP/GR−GCE at various pH and at scan rate 20 mV s^{-1}.

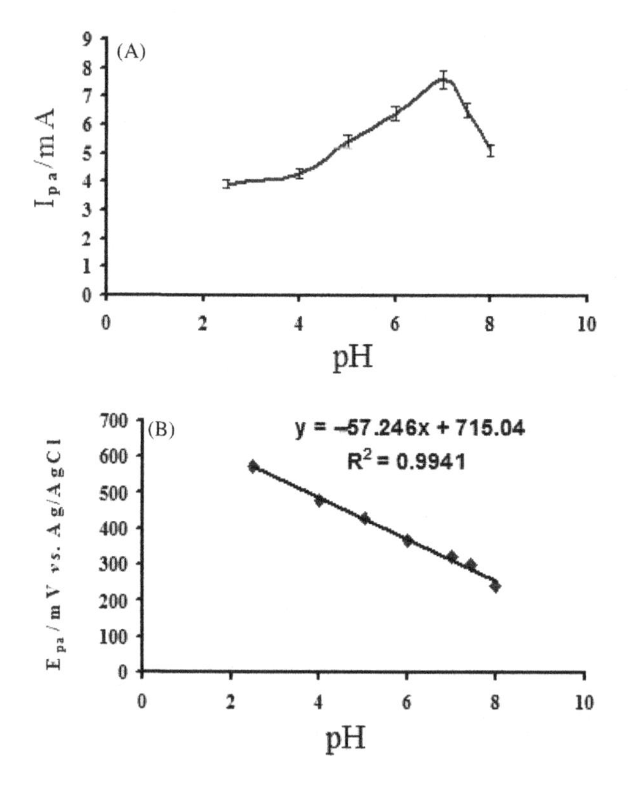

Figure 3.45 Effect of the pH on the (A) anodic peak currents and (B) anodic peak potential.

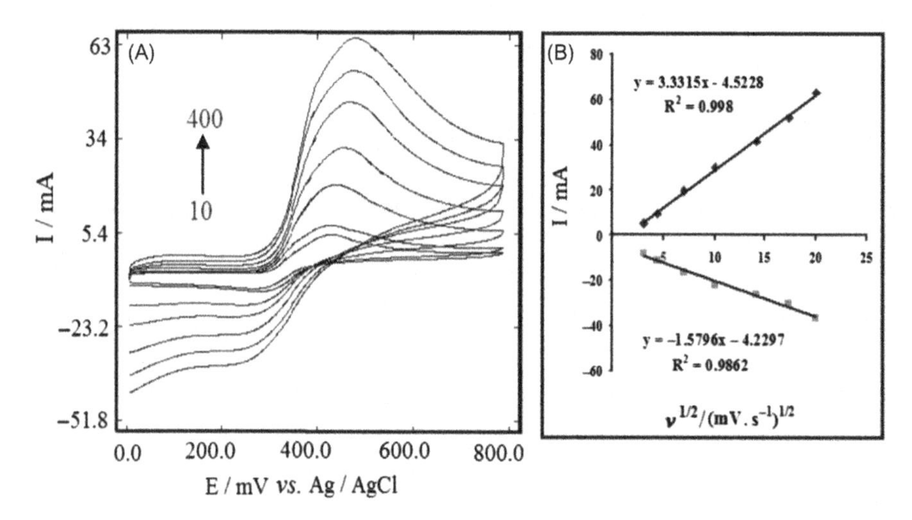

Figure 3.46 (A) CVs of 100 μM PCT in 0.1 M phosphate buffer (pH 7) at P4VP/GR–GCE at scan rates 10, 20, 50, 100, 200, 300, 400 $mV\,s^{-1}$ and (B) Linear relationships of anodic and cathodic peaks currents for 100 μM PCT versus square root of scan rates.

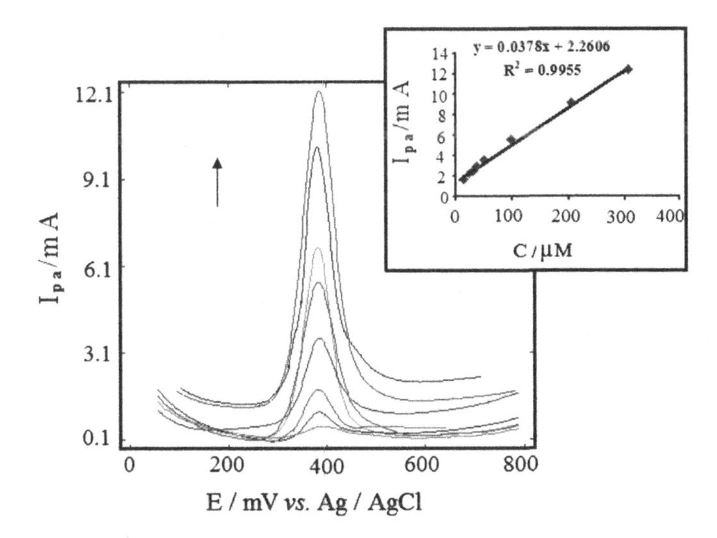

Figure 3.47 DPVs of PCT in 0.1 M phosphate buffer (pH 7) 0.04, 2, 5, 20, 50, 100, 200, and 300 μM at P4VP/GR–GCE. Inset: The corresponding calibration plot.

Table 3.15 Response of Some PCT Sensors Constructed from Various Materials

Electrode Materials	Technique	Detection Limit	Linear Range (μM)	Ref.
Graphene	DPV	0.032 μM	0.1–20	[66]
Polyaniline–MWCNT	SWV	0.25 μM	1–2000	[307]
Nanoparticles Bi_2O_3/GC	CV	0.2 μM	5–1500	[308]
PVC/TTF-TCNQ modified with gold nanoparticles	Amperometric determination	0.66 μM	1–800	[309]
Poly(thaurine)/MWCNT-modified GCE	DPV	0.5 μM	1–100	[242]
MWCNT-alumina-coated silica nanocomposite-modified electrode	SWVs	0.05 μM	0.05–2	[298]
Carbon paste electrode modified with CNT/poly(3-aminophenol)	SWVs	1.1 μM	10–100	[305]
P4VP/GR–GCE	DPV	3.2 nM	0.04–300	Present study

2.0 mV, modulation amplitude of 50 mV and a scan rate of 10 mV s^{-1} (Fig. 3.47). The I_{pa} is linearly proportional to the concentration of PCT in the range of 0.04 to 300 μM. A linear equation of I_{pa} (mA) = 0.0378[PCT] (μM) + 2.260 with (R^2 = 0.995) was obtained. The detection limit of PCT at P4VP/GR–GCE is 3.2 nM, which revealed its better sensitivity than other reported electrodes (Table 3.15).

3.27 DETERMINATION OF PCT IN PHARMACEUTICAL AND BIOLOGICAL SAMPLES

3.27.1 Determination of PCT in Formulation Tablets

P4VP/GR–GCE is tested for the determination of PCT in tablets. The recovery was obtained by using DPV to evaluate the accuracy of the method. The RSD of this method, based on three replicates ($n = 3$), is presented in Table 3.16. Recoveries of PCT at P4VP/GR–GCE in the range of 0.04 to 300 μM reveal that this method is effective and reliable. These findings indicate a successful application of the proposed method for determination of PCT in commercial pharmaceutical formulations.

3.27.2 Determination of PCT in Urine Samples

P4VP/GR–GCE was investigated for the measurement of PCT in three human urine samples. The percentage of recovery of the spiked sample is in the range between 99.41 and 101.01 (Table 3.17). The results show that the developed electrode is suitable for the determination of PCT in biological fluids.

Table 3.16 Determination of PCT in Formulation Tablets Using P4VP/GR–GCE ($n = 3$)

Sample	PCT (μM)			RSD (%)	Recovery (%)
	Contents	Spike	Found		
1	59.2	0	60.3	2.4	101.8
2	59.2	10	68.1	1.1	98.4
3	59.2	20	79.2	1.9	99.9

Table 3.17 Determination of PCT in Human Urine Samples Using P4VP/GR–GCE ($n = 3$)

Sample	PCT (μM)			RSD (%)	Recovery (%)
	Detected	Spike	Found		
1	98.94	25	125.20	1.1	101.0
2	104.23	25	128.40	1.6	99.3
3	112.16	25	136.36	1.5	99.4

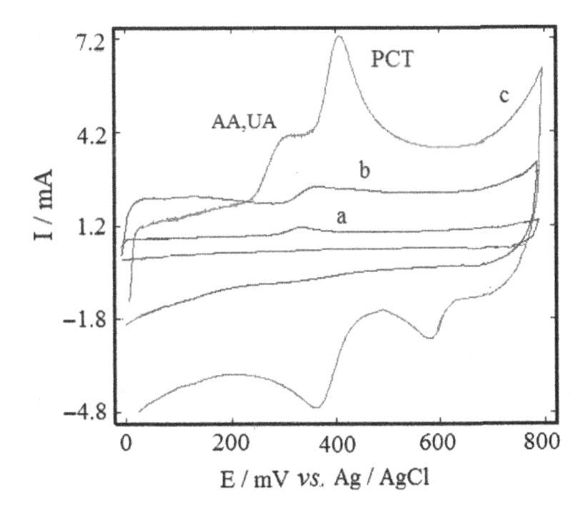

Figure 3.48 CVs of 100 μM AA, 100 μM UA, and 100 μM PCT at the (a) bare GCE, (b) GR−GCE, and (c) P4VP/GR−GCE in 0.1 M phosphate buffer (pH 7) at scan rate 20 mV s^{-1}.

3.28 REPRODUCIBILITY AND STABILITY OF P4VP/GR−GCE

The electrode reproducibility was examined following the same procedure described in Section 3.23. A RSD of the I_{pa} is 3.88% which indicates good reproducibility. The stability of the working electrode was checked for 2 months. It has retained 99% of its current response to the PCT.

3.29 INTERFERENCE STUDIES

UA and AA are the most common constituents found with PCT [267]. The interference of UA and AA on the measurement of PCT on bare GCE, GR−GCE, and P4VP/GR−GCE were studied using CVs. Fig. 3.48a and b indicates that the oxidation peaks of PCT, UA, and AA cannot be separated on the bare GCE and GR−GCE. At the P4VP/GR−GCE, the E_{pa} of AA is close to that of UA at the same potential range. The well-defined wave of PCT was obtained at the P4VP/GR−GCE with good separation from AA and UA (Fig. 3.48c). The anodic peak potentials of AA, UA, and PCT were 300, 310, and 441 mV, respectively. This shows that the voltammetric determination of PCT in biological samples is devoid of any interference from AA and UA.

3.30 CV OF ASA AT THE P4VP/MWCNT−GCE

To investigate the electrocatalytic activity of the P4VP/MWCNT−GCE toward the electrochemical oxidation of ASA, the P4VP/MWCNT−GCE

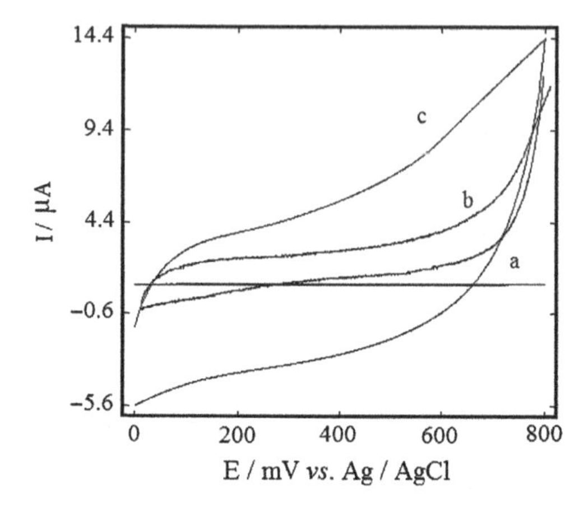

Figure 3.49 CVs of (a) bare GCE, (b) MWCNT−GCE, and (c) P4VP/MWCNT−GCE in 0.1 M phosphate buffer (pH 7.4) at scan rate 20 mV s^{-1}.

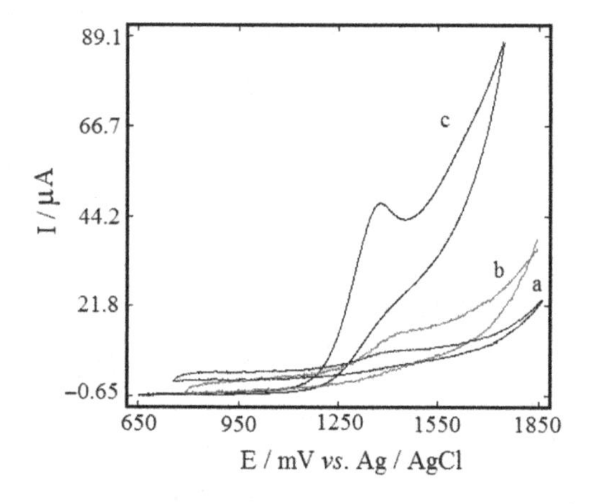

Figure 3.50 CVs of 100 μM ASA at (a) bare GCE, (b) MWCNT−GCE, and (c) P4VP/MWCNT−GCE in 0.1 M phosphate buffer solution (pH 7.4) at scan rate 20 mV s^{-1}.

was characterized by CV ranging from 0.0 to 0.8 V in a 0.1 M phosphate buffer solution (Na_2HPO_4, NaH_2PO_4). The CV obtained at the bare GCE, MWCNT−GCE, and P4VP/MWCNT−GCE in the absence (Fig. 3.49) and in the presence (Fig. 3.50) of 100 μM ASA in 0.1 M phosphate buffer solution is shown.

It's clear from the CV at the bare GCE, MWCNT−GCE, and P4VP/MWCNT−GCE, more noticeable current was observed at the P4VP/MWCNT−GCE. This is attributed to the high surface area of

P4VP/MWCNT−GCE. It can be seen that this modified electrode has a considerable effect on the peak current response of ASA with improvement of more than 30 fold than nonmodified electrode. For the electrochemical oxidation of ASA, the P4VP/MWCNT−GCE also exhibits a significant negative shift of the E_{pa}. The electrochemical oxidation of ASA evident from the oxidation peak potential of ASA are 1400, 1400, and 1281 V for bare GCE, MWCNT−GCE, and P4VP/MWCNT−GCE, respectively. Moreover, the CV response for ASA on bare GCE and MWCNT−GCE are very similar. The negative shift of E_{pa} and the enhancement in I_{pa} response to ASA is sign of electrocatalysis which among others due to increment in the electrode surface area once modified with P4VP and MWCNT.

3.31 CV OF CAFFEINE AT THE P4VP/MWCNT−GCE

CV obtained for 100 μM CF at the bare GCE, MWCNT−GCE, and P4VP/MWCNT−GCE in phosphate buffer solution (pH 7.4) is shown in Fig. 3.51. With addition of 100 μM CF to phosphate buffer, only an oxidation peak appears indicating that the oxidation of CF is an irreversible process [262]. Bare GCE has no response to CF, whereas MWCNT−GCE shows a lower electrochemical response at E_{pa} (1600 mV) (Fig. 3.51a and b) indicating that a sluggish rate of electron transfer. However, an increase in I_{pa} current was observed for CF at

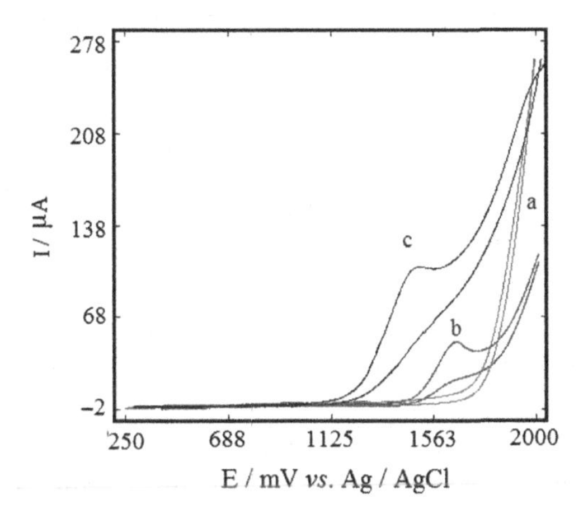

Figure 3.51 CVs of 100 μM caffeine at (a) bare GCE, (b) MWCNT−GCE, and (c) P4VP/MWCNT−GCE in 0.1 M phosphate buffer solution (pH 7.4) at scan rate 20 mV s^{-1}.

P4VP/MWCNT−GCE at 1500 mV indicating that the response sensitivity and stability of CF were greatly improved using the modified electrode. The P4VP/MWCNT−GCE shows negative shift of peak potential when compared to bare GCE and MWCNT−GCE, for the same reason as in Section 3.30.

3.31.1 Effects of pH and Scan Rate on the Oxidation of Caffeine

The electrochemical response of ASA and CF were studied in various supporting electrolytes, viz. phosphate buffer mixtures of KH_2PO_4, K_2HPO_4, Na_2HPO_4, NaH_2PO_4, acetate buffer and H_2SO_4. The best results were obtained using phosphate buffer, where CF showed a sharp oxidation peak, while in the other buffers, broad peaks were obtained. So phosphate buffer was selected for further studies [265]. Fig. 3.52 shows the effect of solution pH on the electrochemical behavior of 100 μM CF in the range of pH 3 to 8. It can be seen that the I_{pa} of CF increases with solution pH until the pH reaches 7.4. However, a lower current was detected when the pH solution was lower or higher than 7.4, suggesting a kinetically less favorable reaction at lower and higher pH.

CVs obtained for 100 μM CF at P4VP/MWCNT−GCE in 0.1 M phosphate buffer solution at different scan rate from 10 to 300 is shown in Fig. 3.53. The I_{pa} of CF increases with increasing scan rates. A good linear relationship between (I_{pa}) and the $v^{1/2}$ were plotted (Fig. 3.53 inset) with the regression equation as: $I_{pa}/\mu A = 6.74\ v^{1/2} + 90.16$ ($R^2 = 0.990$). The I_{pa} were proportional to the $v^{1/2}$ which indicates that the electron transfer reaction is diffusion controlled.

3.32 DETERMINATION OF ASA AND CAFFEINE INDIVIDUALLY

In the individual DPV of various concentrations of ASA (Fig. 3.54A) and CF (Fig. 3.54B) in 0.1 M phosphate buffer (pH 7.4), the I_{pa} has a linear relationship with the concentration of ASA and CF in the range of 0.04 to 350 and 2 to 200, respectively. Two linear equations, i.e., I_{pa} (μA) = 0.2759 [ASA] (μM) + 19.614 and I_{pa} (μA) = 1.0261 [CF] (μM) + 19.295 with linear regression coefficients (R^2) of 0.997 and 0.996, respectively, are obtained. The detection limit of ASA and CF are 4.4 nM and 1.2 nM, respectively.

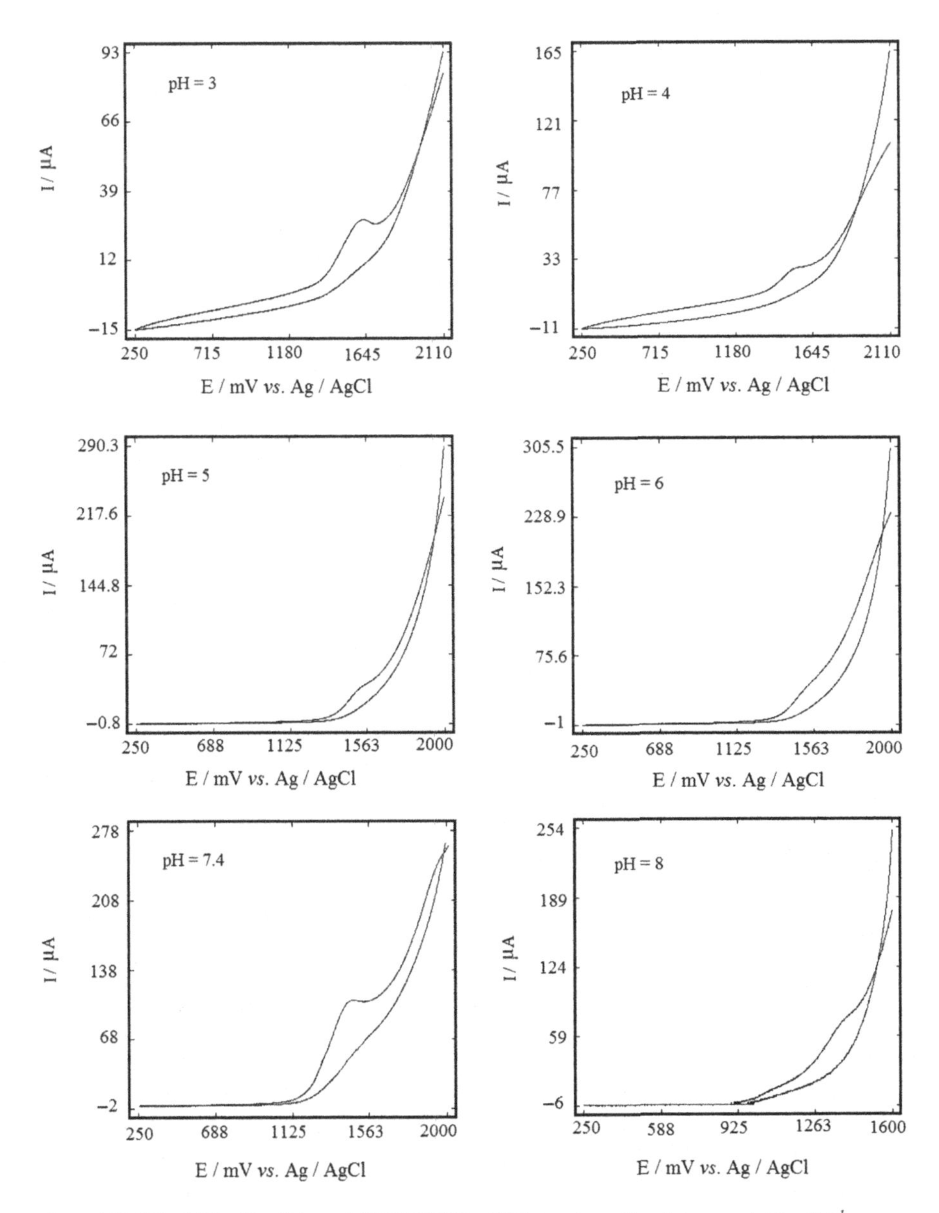

Figure 3.52 CVs of 100 μM caffeine at P4VP/MWCNT−GCE at various pH and at scan rate 20 mV s^{-1}.

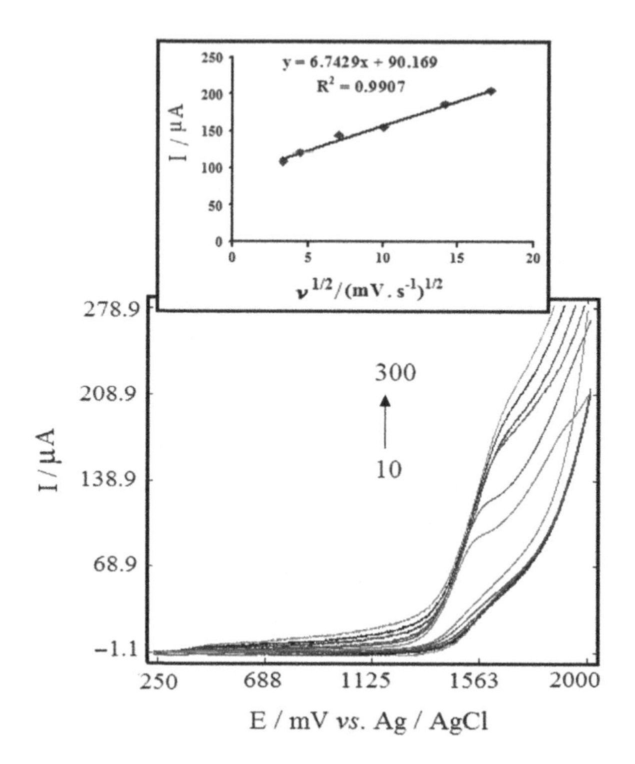

Figure 3.53 CVs of 100 μM caffeine in 0.1 M phosphate buffer (pH 7.4) at P4VP/MWCNT−GCE at scan rates 10, 20, 50, 100, 200, 300 mV s^{-1}. Inset: The relationship of I_{pa} of 100 μM caffeine versus square root of scan rate.

3.33 ANALYTICAL APPLICATIONS

To verify the applicability of the proposed electrochemical sensor, the P4VP/MWCNT−GCE was applied to the determination of ASA in real pharmaceutical samples with DPV. The results are shown in Table 3.18. The recovery was obtained by using DPV to evaluate the accuracy of the method. Satisfactory recoveries of ASA at P4VP/MWCNT−GCE in the range of 0.04 to 350 μM revealed that this method is efficient and reliable. The RSD of this method, based on three replicates ($n = 3$), is presented. Thus, the modified electrode can be recommended for sensitive determination of ASA in tablets.

3.34 SIMULTANEOUS DETERMINATION OF PCT, ASA, AND CAFFEINE

The effect of PCT and CF as the most common interfering substances in the electrochemical determination of ASA was examined by CV technique.

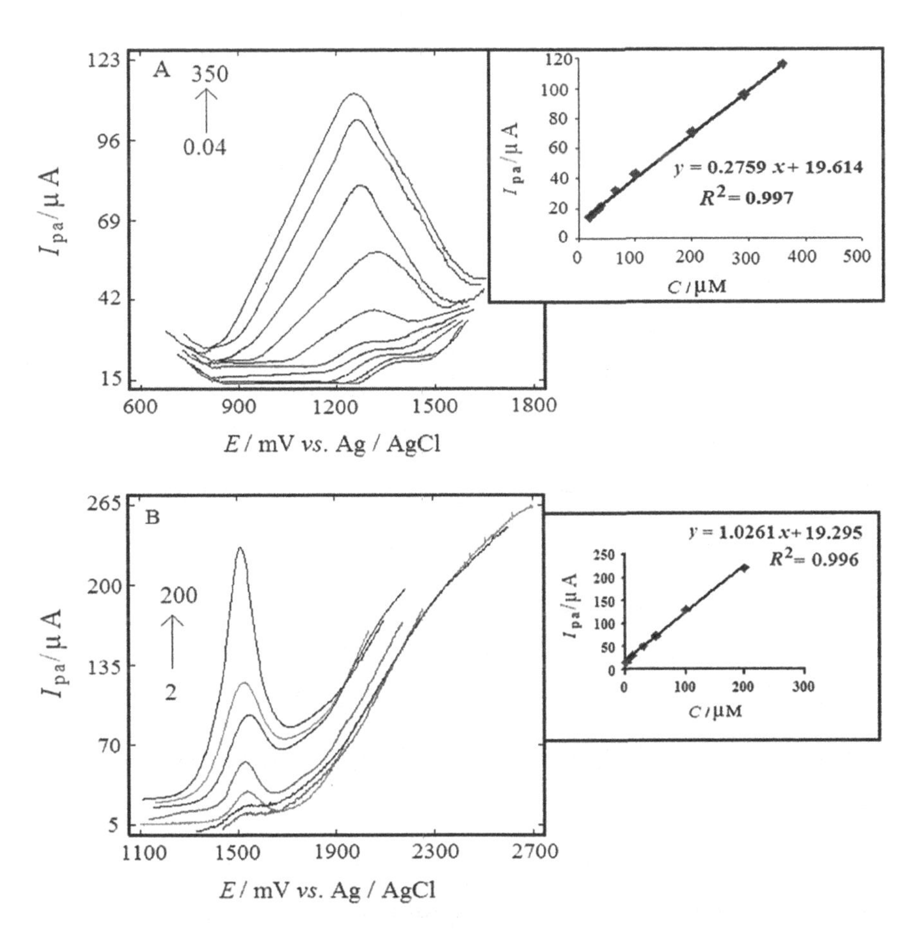

Figure 3.54 DPVs of (A) 0.04, 2, 5, 10, 20, 50, 100, 200, 300, 350 µM ASA; (B) 2, 5, 10, 30, 50, 100, 200 µM CF in phosphate buffer (pH 7.4) at P4VP/MWCNT–GCE.

Table 3.18 Determination of ASA in Formulation Tablets Using P4VP/MWCNT–GCE ($n = 3$)

Sample	ASA (µM)			RSD (%)	Recovery (%)
	Contents	Spike	Found		
1	52.36	0	53.12	1.5	101.4
2	52.36	10	61.50	2.1	98.6
3	52.36	20	72.00	2.7	99.5

Since ASA and CF are two active ingredients in a specific formulation, thus interference of each analyte in the simultaneous analysis of the pairs was investigated.

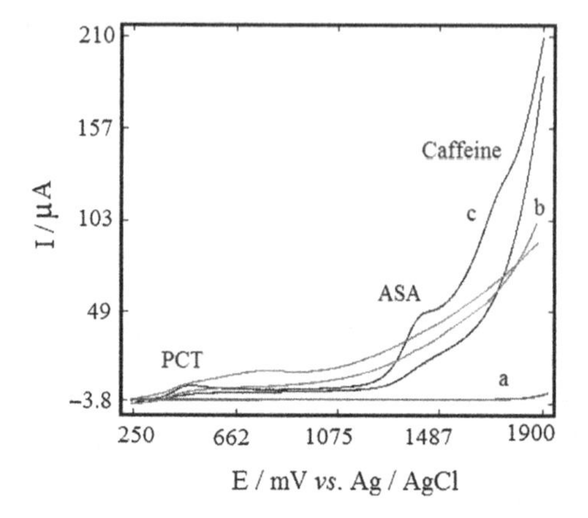

Figure 3.55 CVs of 100 μM PCT, 100 μM ASA and 100 μM caffeine in 0.1 M phosphate buffer solution (pH 7.4) at (a) bare GCE, (b) MWCNT–GCE, and (c) P4VP/MWCNT–GCE at scan rate 20 mV s^{-1}.

A typical CV for a solution having the mixture of ASA, CF, and PCT is shown in Fig. 3.55. The obtained results indicate that the simultaneous determination of the three compounds based on P4VP/MWCNT–GCE is feasible but not on the bare GCE and MWCNT–GCE. This shows that the voltammetric determination of ASA in biological samples is devoid of any interferences from PCT and CF.

3.35 REPRODUCIBILITY AND STABILITY

To ascertain the reproducibility of the electrode, seven electrodes were constructed by the same procedure [310]. The RSD of the I_{pa} was calculated, and it was found to be 3.85%, which indicated that this method gives a good reproducibility for the obtained results. This also confirmed that there was no fouling of reduction of analyte on the electrode.

The stability of the P4VP/MWCNT–GCE was tested by storing the electrode in 0.1 M phosphate buffer (pH 7.4). The operational stability of the electrode remained at 99% of the initial current with constant use for 2 months.

CHAPTER 4

Conclusion

Anodic oxidation of Catechol (CT) in the presence of Thiosemicarbazide (TSC) has been investigated by CV and controlled-potential coulometry. The results show that CT is oxidized to quinone. Quinone takes part in a Michael addition reaction with TSC. The results also show that quinoid can be used as a mediator in the determination of TSC. The proposed method indicates that a cost effective and less labor intensive analysis is possible. The proposed method is considered as a selective method for the voltammetric detection of TSC. The method also can be applied for accurate determination of TSC in pharmaceutical preparations and water samples. Two notable modified electrodes have also been proposed and studied in this research namely P4VP/MWCNT−GCE and P4VP/GR−GCE.

A new electrode based on P4VP/MWCNT−GCE was prepared, characterized, and utilized for the simultaneous quantitative detection of CT and HQ. Both ΔE_p and EIS studies indicate that the kinetics of electron transfer have been enhanced when P4VP was present in the P4VP/MWCNT−GCE as compared to the MWCNT−GCE and bare GCE. The significant increase in peak current together with the sharpness of the peak was observed at the P4VP/MWCNT−GCE, which clearly demonstrated that the P4VP/MWCNT composite is efficient mediators in enhancing the heterogeneous electron transfer for the electrode process of CT and HQ.

In addition, a convenient and simple electrochemical method for the simultaneous quantitative detection of CT and HQ has been developed based on P4VP/GR−GCE. The nanocomposite modified electrode is easily prepared. Incorporation of P4VP with electron mediator property along with GR nanosheet with remarkable conductivity and effective electroactive surface area in this composite mixture has definitely improved the sensitivity of GR−GCE toward HQ and CT. The results indicate that the P4VP/GR nanocomposite can provide a

Electrochemistry of Dihydroxybenzene Compounds. DOI: http://dx.doi.org/10.1016/B978-0-12-813222-7.00004-8

favorable microenvironment to enhance the electron transfer for the electrochemical reaction of CT and HQ. The well-resolved individual anodic peaks of each HQ and CT at P4VP/GR–GCE show that a simultaneous determination of both is possible.

These two developed electrodes are capable for quantitative determination of HQ and CT separately as well as in a mixture as commonly found in water samples. The activities of developed electrodes are not interfered by common coexisting phenolic compounds and also common cationic species in aqueous solution.

Another most important finding to emerge from this study is the fabricated P4VP/MWCNT–GCE and P4VP/GR–GCE which can also be utilized as novel voltammetric sensors for PCT. The proposed electrodes benefited from the synergistic effects of P4VP and MWCNT and also P4VP and GR on the electrooxidation of PCT. The electrodes have a higher electrochemical activity toward the oxidation of PCT as compared with that of either GCE or MWCNT–GCE and GR or GR–GCE. In addition, these two modified electrodes have exhibited a low detection limit, stability, and reproducibility toward PCT without any interference from common physiological interferences i.e., AA and UA. The proposed method has been successfully applied for the determination of PCT in human urine and in pharmaceutical tablets with good precision and accuracy. Moreover, the suggested modified electrodes are viable to be used in analysis of various pharmaceutical compounds such as, ASA, caffeine with satisfactory results.

In conclusion, the high sensitivity, stability and reproducibility, simple fabrication procedures, and lower detection limits are among the advantages of these two modified electrodes. The modified electrodes in this work could be used as sensors for highly sensitive, for catalytic oxidation, and simultaneous determination of pharmaceutically important compounds and diphenols.

REFERENCES

[1] Romig Jr AD, Baker AB, Johannes J, Zipperian T, Eijkel K, Kirchhoff B, et al. An introduction to nanotechnology policy: opportunities and constraints for emerging and established economies. Technol Forecast Soc Change 2007;74(9):1634–42.

[2] Zitt M, Bassecoulard E. Delineating complex scientific fields by an hybrid lexical-citation method: an application to nanosciences. Inf Process Manage 2006;42(6):1513–31.

[3] Russell R, Cresanti R. Technical Report. Research and development leading to a revolution in technology and industry. National Nanotechnology Coordination Office; 2007. p. 1–41.

[4] Rao CNR, Muller A, Cheetham AK, editors. The chemistry of nanomaterials, 1. Weinheim: Wiley-VCH, Verlag GmbH & Co. KGaA; 2006. p. xvi. 113, 698 and 702.

[5] Rao CNR, Cheetham AK. Science and technology of nanomaterials: current status and future prospects. J Mater Chem 2001;11(12):2887–94.

[6] Cao G. Nanostructures & nanomaterials: synthesis, properties & applications. London: Imperial College Press; 2004. p. 2–4, 93–95, 97–104, 144, 161, 173–176, 182, 189, 218 and 219.

[7] Ogawa H, Nishikawa M, Abe A. Hall measurement studies and an electrical conduction model of tin oxide ultrafine particle films. J Appl Phys 1982;53(6):4448–55.

[8] Luth H. Surfaces and interfaces of solid materials. 3rd ed. Berlin: Springer; 1995. p. 21.

[9] Murty B, Shankar P, Raj B, Rath B, Murday J. The big world of nanomaterials. Textbook of nanoscience and nanotechnology. Universities Press (India) Private Limited; 2013. p. 1–28, Springer: Berlin-Heidelberg: Berlin, Germany.

[10] Gong ZQ. Electrochemical characterization of various modified composite pencil graphite electrodes. PhD Thesis. Universiti Sains Malaysia; 2012. p. 2–13.

[11] Mauter MS, Elimelech M. Environmental applications of carbon-based nanomaterials. Environ Sci Technol 2008;42(16):5843–59.

[12] <http://simple.wikipedia.org/wiki/Carbon_nanotube>, accessed on 31.5.2013.

[13] <http://ipn2.epfl.ch/CHBU/images/rollup.gif>, accessed on 31.5.2013.

[14] Valcarcel M, Simonet BM, Cardenas S, Suarez B. Present and future applications of carbon nanotubes to analytical science. Anal Bioanal Chem 2005;382(8):1783–90.

[15] Trojanowicz M. Analytical applications of carbon nanotubes: a review. Trends Anal Chem 2006;25(5):480–9.

[16] Merkoci A. Carbon nanotubes in analytical sciences. Microchim Acta 2006;152 (3–4):157–74.

[17] Wildgoose GG, Banks CE, Leventis HC, Compton RG. Chemically modified carbon nanotubes for use in electroanalysis. Microchim Acta 2006;152(3–4):187–214.

[18] Merkoci A, Pumera M, Llopis X, Pérez B, del Valle M, Alegret S. New materials for electrochemical sensing VI: carbon nanotubes. Trends Anal Chem 2005;24(9):826–38.

[19] Wang J. Nanomaterial-based electrochemical biosensors. Analyst 2005;130(4):421–6.

[20] Nugent J, Santhanam K, Rubio A, Ajayan P. Fast electron transfer kinetics on multiwalled carbon nanotube microbundle electrodes. Nano Lett 2001;1(2):87–91.

[21] Wang H-S, Li T-H, Jia W-L, Xu H-Y. Highly selective and sensitive determination of dopamine using a Nafion/carbon nanotubes coated poly(3-methylthiophene) modified electrode. Biosens Bioelectron 2006;22(5):664–9.

[22] Zhang M, Liu K, Xiang L, Lin Y, Su L, Mao L. Carbon nanotube-modified carbon fiber microelectrodes for in vivo voltammetric measurement of ascorbic acid in rat brain. Anal Chem 2007;79(17):6559–65.

[23] Lu TL, Tsai YC. Electrocatalytic oxidation of acetylsalicylic acid at multiwalled carbon nanotube-alumina-coated silica nanocomposite modified glassy carbon electrodes. Sens. Actuators, B: Chem 2010;148(2):590–4.

[24] Beitollahi H, Mohadesi A, Mohammadi S, Akbari A. Electrochemical behavior of a carbon paste electrode modified with 5-amino-3′,4′-dimethyl-biphenyl-2-ol/carbon nanotube and its application for simultaneous determination of isoproterenol, acetaminophen and *N*-acetylcysteine. Electrochim Acta 2012;68(0):220–6.

[25] Profumo A, Fagnoni M, Merli D, Quartarone E, Protti S, Dondi D, et al. Multiwalled carbon nanotube chemically modified gold electrode for inorganic as speciation and Bi (III) determination. Anal Chem 2006;78(12):4194–9.

[26] Tsai YC, Chen JM, Marken F. Simple cast-deposited multi-walled carbon nanotube/Nafion™ thin film electrodes for electrochemical stripping analysis. Microchim Acta 2005;150(3):269–76.

[27] Li Y, Tang L, Li J. Preparation and electrochemical performance for methanol oxidation of pt/graphene nanocomposites. Electrochem Commun 2009;11(4):846–9.

[28] Liu S, Tian J, Wang L, Luo Y, Lu W, Sun X. Self-assembled graphene platelet–glucose oxidase nanostructures for glucose biosensing. Biosens Bioelectron 2011;26(11):4491–6.

[29] Wu Z-S, Pei S, Ren W, Tang D, Gao L, Liu B, et al. Field emission of single-layer graphene films prepared by electrophoretic deposition. Adv Mater 2009;21(17):1756–60.

[30] Novoselov K, Geim A, Morozov S, Jiang D, Zhang Y, Dubonos S, et al. Electric field effect in atomically thin carbon films. Science 2004;306(5696):666–9.

[31] Li D, Kaner RB. Graphene-based materials. Nat Nanotechnol 2008;3:101.

[32] Geim AK, Novoselov KS. The rise of graphene. Nat Mater 2007;6(3):183–91.

[33] Ao ZM, Yang J, Li S, Jiang Q. Enhancement of CO detection in Al doped graphene. Chem Phys Lett 2008;461(4–6):276–9.

[34] Vivekchand S, Rout CS, Subrahmanyam K, Govindaraj A, Rao C. Graphene-based electrochemical supercapacitors. J Chem Sci 2008;120(1):9–13.

[35] Zhu Y, Murali S, Cai W, Li X, Suk JW, Potts JR, et al. Graphene and graphene oxide: synthesis, properties, and applications. Adv Mater 2010;22(35):3906–24.

[36] Liu F, Choi JY, Seo TS. Graphene oxide arrays for detecting specific DNA hybridization by fluorescence resonance energy transfer. Biosens Bioelectron 2010;25(10):2361–5.

[37] Cui F, Zhang X. Electrochemical sensor for epinephrine based on a glassy carbon electrode modified with graphene/gold nanocomposites. J Electroanal Chem 2012;669:35–41.

[38] Shao Y, Wang J, Wu H, Liu J, Aksay IA, Lin Y. Graphene based electrochemical sensors and biosensors: a review. Electroanalysis 2010;22(10):1027–36.

[39] Wang Y, Li Y, Tang L, Lu J, Li J. Application of graphene-modified electrode for selective detection of dopamine. Electrochem Commun 2009;11(4):889–92.

[40] Yang W, Ratinac KR, Ringer SP, Thordarson P, Gooding JJ, Braet F. Carbon nanomaterials in biosensors: should you use nanotubes or graphene? Angew Chem Int Ed 2010;49 (12):2114–38.

[41] Jung JH, Cheon DS, Liu F, Lee KB, Seo TS. A graphene oxide based immuno-biosensor for pathogen detection. Angew Chem Int Ed 2010;49(33):5708–11.

[42] Mallesha M, Manjunatha R, Nethravathi C, Suresh GS, Rajamathi M, Melo JS, et al. Functionalized-graphene modified graphite electrode for the selective determination of dopamine in presence of uric acid and ascorbic acid. Bioelectrochemistry 2011;81(2):104–8.

[43] Wu H, Wang J, Kang X, Wang C, Wang D, Liu J, et al. Glucose biosensor based on immobilization of glucose oxidase in platinum nanoparticles/graphene/chitosan nanocomposite film. Talanta 2009;80(1):403–6.

[44] Shan C, Yang H, Song J, Han D, Ivaska A, Niu L. Direct electrochemistry of glucose oxidase and biosensing for glucose based on graphene. Anal Chem 2009;81(6):2378–82.

[45] Shan C, Yang H, Han D, Zhang Q, Ivaska A, Niu L. Graphene/AuNPs/chitosan nanocomposites film for glucose biosensing. Biosens Bioelectron 2010;25(5):1070.

[46] Kang X, Wang J, Wu H, Aksay IA, Liu J, Lin Y. Glucose oxidase–graphene–chitosan modified electrode for direct electrochemistry and glucose sensing. Biosens Bioelectron 2009;25(4):901–5.

[47] Lin WJ, Liao CS, Jhang JH, Tsai YC. Graphene modified basal and edge plane pyrolytic graphite electrodes for electrocatalytic oxidation of hydrogen peroxide and β-nicotinamide adenine dinucleotide. Electrochem Commun 2009;11(11):2153–6.

[48] Liu H, Gao J, Xue M, Zhu N, Zhang M, Cao T. Processing of graphene for electrochemical application: noncovalently functionalize graphene sheets with water-soluble electroactive methylene green. Langmuir 2009;25(20):12006–10.

[49] Zhou M, Zhai Y, Dong S. Electrochemical sensing and biosensing platform based on chemically reduced graphene oxide. Anal Chem 2009;81(14):5603–13.

[50] Guo HL, Wang XF, Qian QY, Wang FB, Xia XH. A green approach to the synthesis of graphene nanosheets. ACS Nano 2009;3(9):2653–9.

[51] Shao Y, Wang J, Engelhard M, Wang C, Lin Y. Facile and controllable electrochemical reduction of graphene oxide and its applications. J Mater Chem 2010;20(4):743–8.

[52] Zhou M, Wang Y, Zhai Y, Zhai J, Ren W, Wang F, et al. Controlled synthesis of large-area and patterned electrochemically reduced graphene oxide films. Chem Eur J 2009;15 (25):6116–20.

[53] <http://origin-ars.els-cdn.com/content/image/1-s2.0-S0079642511000442-gr6.jpg>, accessed on 31.5.2013.

[54] Wang X, Zhi L, Müllen K. Transparent, conductive graphene electrodes for dye-sensitized solar cells. Nano Lett 2008;8(1):323–7.

[55] Eda G, Fanchini G, Chhowalla M. Large-area ultrathin films of reduced graphene oxide as a transparent and flexible electronic material. Nat Nanotechnol 2008;3(5):270–4.

[56] Li D, Müller MB, Gilje S, Kaner RB, Wallace GG. Processable aqueous dispersions of graphene nanosheets. Nat Nanotechnol 2008;3(2):101–5.

[57] Barth, A. and Marx, W., Graphene – a rising star in view of scientometrics. arXiv preprint arXiv:0808.3320, 2008.

[58] Segal M. Selling graphene by the ton. Nat Nanotechnol 2009;4(10):612–14.

[59] Viculis LM, Mack JJ, Mayer OM, Hahn HT, Kaner RB. Intercalation and exfoliation routes to graphite nanoplatelets. J Mater Chem 2005;15(9):974–8.

[60] Afanasov IM, Morozov VA, Kepman A, Ionov S, Seleznev A, Tendeloo GV, et al. Preparation, electrical and thermal properties of new exfoliated graphite-based composites. Carbon NY 2009;47(1):263–70.

[61] Wallace P. The band theory of graphite. Phys Rev 1947;71(9):622.

[62] Niyogi S, Bekyarova E, Itkis ME, McWilliams JL, Hamon MA, Haddon RC. Solution properties of graphite and graphene. J Am Chem Soc 2006;128(24):7720–1.

[63] Shan C, Yang H, Han D, Zhang Q, Ivaska A, Niu L. Water-soluble graphene covalently functionalized by biocompatible poly-L-lysine. Langmuir 2009;25(20):12030–3.

[64] Si Y, Samulski ET. Synthesis of water soluble graphene. Nano Lett 2008;8(6):1679–82.

[65] Tang Z, Wu H, Cort JR, Buchko GW, Zhang Y, Shao Y, et al. Constraint of DNA on functionalized graphene improves its biostability and specificity. Small 2010;6(11):1205–9.

[66] Kang X, Wang J, Wu H, Liu J, Aksay IA, Lin Y. A graphene-based electrochemical sensor for sensitive detection of paracetamol. Talanta 2010;81(3):754–9.

[67] Kuila T, Bose S, Mishra AK, Khanra P, Kim NH, Lee NH. Chemical functionalization of graphene and its applications. Prog Mater Sci 2012;1061–105.

[68] Kaniyoor A, Ramaprabhu S. Soft functionalization of graphene for enhanced tri-iodide reduction in dye sensitized solar cells. J Mater Chem 2012;22(17):8377–84.

[69] Liu Z, Wang Z, Cao Y, Jing Y, Liu Y. High sensitive simultaneous determination of hydroquinone and catechol based on graphene/$BMIMPF_6$ nanocomposite modified electrode. Sens. Actuators, B: Chem 2011;157(2):540–6.

[70] Wang X, Zhi L, Tsao N, TomovićÂ Ž, Li J, Müllen K. Transparent carbon films as electrodes in organic solar cells. Angew Chem 2008;120(16):3032–4.

[71] Yoo EJ, Kim J, Hosono E, Zhou H, Kudo T, Honma I. Large reversible Li storage of graphene nanosheet families for use in rechargeable lithium ion batteries. Nano Lett 2008;8 (8):2277–82.

[72] Bong S, Kim YR, Kim I, Woo S, Uhm S, Lee J, et al. Graphene supported electrocatalysts for methanol oxidation. Electrochem Commun 2010;12(1):129–31.

[73] Li J, Guo S, Zhai Y, Wang E. High-sensitivity determination of lead and cadmium based on the Nafion-graphene composite film. Anal Chim Acta 2009;649(2):196–201.

[74] Hong W, Bai H, Xu Y, Yao Z, Gu Z, Shi G. Preparation of gold nanoparticle/graphene composites with controlled weight contents and their application in biosensors. J Phys Chem C 2010;114(4):1822–6.

[75] Ren W, Luo HQ, Li NB. Simultaneous voltammetric measurement of ascorbic acid, epinephrine and uric acid at a glassy carbon electrode modified with caffeic acid. Biosens Bioelectron 2006;21(7):1086–92.

[76] Al-Saleh MH, Sundararaj U. Review of the mechanical properties of carbon nanofiber/polymer composites. Composites A: Appl Sci Manuf 2011;42(12):2126–42.

[77] Denault J, Labrecque B. National Research Council Canada, 75 de Mortagne Blvd. Boucherville, Québec, J4B 6Y4 Technology group on polymer nanocomposites–PNC-Tech. Industrial Materials Institute; 2004.

[78] Salavagione HJ, Martínez G, Ellis G. Graphene-based polymer nanocomposites. Rijeka: Physics and Applications of Graphene—Experiments, In-Tech; 2011. p. 169–92.

[79] Lagashetty A, Venkataraman A. Polymer nanocomposites. Resonance 2005;10(7):49–57.

[80] Jacobs CB, Peairs MJ, Venton BJ. Review: carbon nanotube based electrochemical sensors for biomolecules. Anal Chim Acta 2010;662(2):105–27.

[81] Gutierrez F, Rubianes MD, Rivas GA. Dispersion of multi-wall carbon nanotubes in glucose oxidase: characterization and analytical applications for glucose biosensing. Sens. Actuators, B: Chem 2012;161(1):191–7.

[82] Valentini F, Amine A, Orlanducci S, Terranova ML, Palleschi G. Carbon nanotube purification: preparation and characterization of carbon nanotube paste electrodes. Anal Chem 2003;75(20):5413–21.

[83] Guo M, Chen J, Nie L, Yao S. Electrostatic assembly of calf thymus DNA on multi-walled carbon nanotube modified gold electrode and its interaction with chlorpromazine hydrochloride. Electrochim Acta 2004;49(16):2637–43.

[84] Luo H, Shi Z, Li N, Gu Z, Zhuang Q. Investigation of the electrochemical and electrocatalytic behavior of single-wall carbon nanotube film on a glassy carbon electrode. Anal Chem 2001;73(5):915–20.

[85] An KH, Jeong SY, Hwang HR, Lee YH. Enhanced sensitivity of a gas sensor incorporating single-walled carbon nanotube–polypyrrole nanocomposites. Adv Mater 2004;16 (12):1005–9.

[86] Wang J, Dai J, Yarlagadda T. Carbon nanotube-conducting-polymer composite nanowires. Langmuir 2005;21(1):9–12.

[87] Sinani VA, Gheith MK, Yaroslavov AA, Rakhnyanskaya AA, Sun K, Mamedov AA, et al. Aqueous dispersions of single-wall and multiwall carbon nanotubes with designed amphiphilic polycations. J Am Chem Soc 2005;127(10):3463–72.

[88] Hong S, Kim M, Hong CK, Jung D, Shim SE. Encapsulation of multi-walled carbon nanotubes by poly(4-vinylpyridine) and its dispersion stability in various solvent media. Synth Met 2008;158(21–24):900–7.

[89] Kharisov BI, Kharissova OV, Leija Gutierrez H, Ortiz Méndez U. Recent advances on the soluble carbon nanotubes. Ind Eng Chem Res 2008;48(2):572–90.

[90] O'connell MJ, Bachilo SM, Huffman CB, Moore VC, Strano MS, Haroz EH, et al. Band gap fluorescence from individual single-walled carbon nanotubes. Science 2002;297 (5581):593–6.

[91] Islam M, Rojas E, Bergey D, Johnson A, Yodh A. High weight fraction surfactant solubilization of single-wall carbon nanotubes in water. Nano Lett 2003;3(2):269–73.

[92] O'Connell MJ, Boul P, Ericson LM, Huffman C, Wang Y, Haroz E, et al. Reversible water-solubilization of single-walled carbon nanotubes by polymer wrapping. Chem Phys Lett 2001;342(3):265–71.

[93] Gomez FJ, Chen RJ, Wang D, Waymouth RM, Dai H. Ring opening metathesis polymerization on non-covalently functionalized single-walled carbon nanotubes. Chemical Communications 2003;0(2):190–1.

[94] Barraza HJ, Pompeo F, O'Rear Edgar A, Resasco DE. SWNT-filled thermoplastic and elastomeric composites prepared by miniemulsion polymerization. Nano Lett 2002;2 (8):797–802.

[95] Star A, Stoddart JF, Steuerman D, Diehl M, Boukai A, Wong EW, et al. Preparation and properties of polymer-wrapped single-walled carbon nanotubes. Angew Chem Int Ed 2001;40(9):1721–5.

[96] Moniruzzaman M, Winey KI. Polymer nanocomposites containing carbon nanotubes. Macromolecules 2006;39(16):5194–205.

[97] Winey KI, Vaia RA. Polymer nanocomposites. Mater Res Soc Bull 2007;32(04):314–22.

[98] Stankovich S, Dikin DA, Dommett GH, Kohlhaas KM, Zimney EJ, Stach EA, et al. Graphene-based composite materials. Nature 2006;442(7100):282–6.

[99] Dreyer DR, Park S, Bielawski CW, Ruoff RS. The chemistry of graphene oxide. Chem Soc Rev 2010;39(1):228–40.

[100] Wang G, Yang J, Park J, Gou X, Wang B, Liu H, et al. Facile synthesis and characterization of graphene nanosheets. J Phys Chem C 2008;112(22):8192–5.

[101] Wang G, Shen X, Wang B, Yao J, Park J. Synthesis and characterisation of hydrophilic and organophilic graphene nanosheets. Carbon NY 2009;47(5):1359–64.

[102] Li X, Wang X, Zhang L, Lee S, Dai H. Chemically derived, ultrasmooth graphene nanoribbon semiconductors. Science 2008;319(5867):1229–32.

[103] Blake P, Brimicombe PD, Nair RR, Booth TJ, Jiang D, Schedin F, et al. Graphene-based liquid crystal device. Nano Lett 2008;8(6):1704–8.

[104] Stoller MD, Park S, Zhu Y, An J, Ruoff RS. Graphene-based ultracapacitors. Nano Lett 2008;8(10):3498–502.

[105] Liu Z, Liu Q, Huang Y, Ma Y, Yin S, Zhang X, et al. Organic photovoltaic devices based on a novel acceptor material: graphene. Adv Mater 2008;20(20):3924–30.

[106] Robinson JT, Perkins FK, Snow ES, Wei Z, Sheehan PE. Reduced graphene oxide molecular sensors. Nano Lett 2008;8(10):3137–40.

[107] Ang PK, Chen W, Wee ATS, Loh KP. Solution-gated epitaxial graphene as pH sensor. J Am Chem Soc 2008;130(44):14392–3.

[108] Wu J, Becerril HA, Bao Z, Liu Z, Chen Y, Peumans P. Organic solar cells with solution-processed graphene transparent electrodes. Appl Phys Lett 2008;92(26) 263302.

[109] Ramanathan T, Abdala A, Stankovich S, Dikin D, Herrera-Alonso M, Piner R, et al. Functionalized graphene sheets for polymer nanocomposites. Nat Nanotechnol 2008;3 (6):327–31.

[110] Allen MJ, Tung VC, Kaner RB. Honeycomb carbon: a review of graphene. Chem Rev 2010;110(1):132.

[111] Aalaie J, Rahmatpour A, Maghami S. Preparation and characterization of linear low density polyethylene/carbon nanotube nanocomposites. J Macromol Sci, B: Phys 2007;46 (5):877–89.

[112] Li L, Li CY, Ni C, Rong L, Hsiao B. Structure and crystallization behavior of Nylon 66/ multi-walled carbon nanotube nanocomposites at low carbon nanotube contents. Polymer (Guildf) 2007;48(12):3452–60.

[113] Morales-Teyssier O, Sánchez-Valdes S, Ramos-de Valle LF. Effect of carbon nanofiber functionalization on the dispersion and physical and mechanical properties of polystyrene nanocomposites. Macromol Mater Eng 2006;291(12):1547–55.

[114] Mrozek RA, Kim B-S, Holmberg VC, Taton TA. Homogeneous, coaxial liquid crystal domain growth from carbon nanotube seeds. Nano Lett 2003;1665–9.

[115] Bliznyuk VN, Singamaneni S, Sanford RL, Chiappetta D, Crooker B, Shibaev PV. Matrix mediated alignment of single wall carbon nanotubes in polymer composite films. Polymer (Guildf) 2006;47(11):3915–21.

[116] Zhao B, Hu H, Haddon RC. Synthesis and properties of a water-soluble single-walled carbon nanotube–poly(*m*-aminobenzene sulfonic acid) graft copolymer. Adv Funct Mater 2004;14(1):71–6.

[117] Ago H, Petritsch K, Shaffer MSP, Windle AH, Friend RH. Composites of carbon nanotubes and conjugated polymers for photovoltaic devices. Adv Mater 1999;11 (15):1281–5.

[118] Chiang C, Fincher Jr C, Park Y, Heeger A, Shirakawa H, Louis E, et al. Electrical conductivity in doped polyacetylene. Phys Rev Lett 1977;39(17):1098–101.

[119] Shirakawa H, Louis EJ, MacDiarmid AG, Chiang CK, Heeger AJ. Synthesis of electrically conducting organic polymers: halogen derivatives of polyacetylene, (CH). J Chem Soc Chem Commun 1977;0(16):578–80.

[120] Ling JLW. Poly(4-vinylpyridine) and its corresponding copolymers based membrane sensor for cadmium(II). MSc Thesis. Universiti Sains Malaysia; June 2012. p. 8.

[121] Zhang L, Zujovic ZD, Peng H, Bowmaker GA, Kilmartin PA, Travas-Sejdic J. Structural characteristics of polyaniline nanotubes synthesized from different buffer solutions. Macromolecules 2008;41(22):8877–84.

[122] Zhang X, Manohar SK. Narrow pore-diameter polypyrrole nanotubes. J Am Chem Soc 2005;127(41):14156–7.

[123] Yang Z, Kou X, Ni W, Sun Z, Li L, Wang J. Fluorescent mesostructured polythiophene–silica composite particles synthesized by in situ polymerization of structure-directing monomers. Chem Mater 2007;19(25):6222–9.

[124] De Rossi D, Carpi F, Scilingo EP. Polymer based interfaces as bioinspired 'smart skins'. Adv Colloid Interface Sci 2005;116(1):165–78.

[125] Li N, Huang Y, Du F, He X, Lin X, Gao H, et al. Electromagnetic interference (EMI) shielding of single-walled carbon nanotube epoxy composites. Nano Lett 2006;6(6):1141–5.

[126] Yang Y, Gupta MC, Dudley KL, Lawrence RW. Novel carbon nanotube-polystyrene foam composites for electromagnetic interference shielding. Nano Lett 2005;5(11):2131–4.

[127] Sein Jr LT. Effect of substituting oxygen for terminal nitrogen in aniline oligomers: a DFT comparison of hydroxyl and amino terminated aniline trimers. J Phys Chem A 2008;112 (12):2598–603.

[128] Skorb EV, Skirtach AG, Sviridov DV, Shchukin DG, Möhwald H. Laser-controllable coatings for corrosion protection. ACS Nano 2009;3(7):1753–60.

[129] Ryu KS, Jeong SK, Joo J, Kim KM. Polyaniline doped with dimethyl sulfate as a nucleophilic dopant and its electrochemical properties as an electrode in a lithium secondary battery and a redox supercapacitor. J Phys Chem B 2007;111(4):731–9.

[130] Marchioni F, Yang J, Walker W, Wudl F. A low band gap conjugated polymer for supercapacitor devices. J Phys Chem B 2006;110(44):22202–6.

[131] Yu L, Yang H, Ai X, Cao Y. Structural and electrochemical characterization of nanocrystalline Li [Li0.12Ni0.32Mn0.56] O_2 synthesized by a polymer–pyrolysis route. J Phys Chem B 2005;109(3):1148–54.

[132] Pelah A, Seemann R, Jovin TM. Reversible cell deformation by a polymeric actuator. J Am Chem Soc 2007;129(3):468–9.

[133] Kim J, Seo YB. Electro-active paper actuators. Smart Mater Struct 2002;11(3):355.

[134] Parsa A. Studies on electrochemically synthesized polyaniline and its copolymers. PhD thesis. Universiti Sains Malaysia; September 2009. p. 1.

[135] Frackowiak E, Khomenko V, Jurewicz K, Lota K, Béguin F. Supercapacitors based on conducting polymers/nanotubes composites. J Power Sources 2006;153(2):413–18.

[136] Lange U, Roznyatovskaya NV, Mirsky VM. Conducting polymers in chemical sensors and arrays. Anal Chim Acta 2008;614(1):1–26.

[137] Guimard NK, Gomez N, Schmidt CE. Conducting polymers in biomedical engineering. Prog Polym Sci 2007;32(8–9):876–921.

[138] Fan Y, Liu J-H, Yang C-P, Yu M, Liu P. Graphene–polyaniline composite film modified electrode for voltammetric determination of 4-aminophenol. Sens. Actuators, B: Chem 2011;157(2):669–74.

[139] Zhuang Z, Li J, Xu R, Xiao D. Electrochemical detection of dopamine in the presence of ascorbic acid using overoxidized polypyrrole/graphene modified electrodes. Int J Electrochem Soc 2011;6:2149–61.

[140] Zeng Z, Gupta S, Huang H, Yeager E. Oxygen reduction on poly(4-vinylpyridine)-modified ordinary pyrolytic graphite electrodes with adsorbed cobalt tetra-sulphonated phthalocyanine in acid solutions. J Appl Electrochem 1991;21(11):973–81.

[141] Wang J, Golden T, Peng T. Poly(4-vinylpyridine)-coated glassy carbon flow detectors. Anal Chem 1987;59(5):740–4.

[142] Li J, Qiu J-D, Xu J-J, Chen H-Y, Xia X-H. The synergistic effect of prussian-blue-grafted carbon nanotube/poly(4-vinylpyridine) composites for amperometric sensing. Adv Funct Mater 2007;17(9):1574–80.

[143] Mamedov AA, Kotov NA, Prato M, Guldi DM, Wicksted JP, Hirsch A. Molecular design of strong single-wall carbon nanotube/polyelectrolyte multilayer composites. Nat Mater 2002;1(3):190–4.

[144] Qin S, Qin D, Ford WT, Herrera JE, Resasco DE. Grafting of poly(4-vinylpyridine) to single-walled carbon nanotubes and assembly of multilayer films. Macromolecules 2004;37 (26):9963–7.

[145] Verdejo R, Bernal MM, Romasanta LJ, Lopez-Manchado MA. Graphene filled polymer nanocomposites. J Mater Chem 2011;21(10):3301–10.

[146] Ahmad F, Chow FC, Ho YW, Chin YS, Christenson A, Bainbridge M, et al. Development of prototype wireless transmission measurement for glucose in subcutaneous and brain striatum. Electroanalysis 2008;20(9):1008–15.

[147] Shono T. Electroorganic synthesis. Kyoto, Japan: Academic Press Limited; 1991. p. 1–5.

[148] Nematollahi D, Amani A. Electrochemical synthesis of the new substituted phenylpiperazines. J Electroanal Chem 2011;651(1):72–9.

[149] Nematollahi D, Esmaili R. A green approach for the electrochemical synthesis of 4-morpholino-2-(arylsulfonyl)benzenamines. Tetrahedron Lett 2010;51(37):4862–5.

[150] Nematollahi D, Bamzadeh M, Shayani-Jam H. Electrochemical oxidation of catechols in the presence of phenyl-meldrum's acid. Synthesis and kinetic evaluation. Chem Pharm Bull 2010;58(1):23–6.

[151] Fakhari AR, Ahmar H, Davarani SSH, Shaabani A, Nikjah S, Maleki A. Electro-organic synthesis of 2-amino-3-cyano-benzofuran derivatives using hydroquinones and malononitrile. Synth Commun 2011;41(4):561–8.

[152] Shahrokhian S, Rastgar S. Investigation of the electrochemical behavior of catechol and 4-methylcatechol in the presence of methyl mercapto thiadiazol as a nucleophile: application to electrochemical synthesis. J Appl Electrochem 2010;40(1):115–22.

[153] Zeng C-C, Ping D-W, Hu L-M, Song X-Q, Zhong R-G. Anodic oxidation of catechols in the presence of α-oxoketene *N*,*N*-acetals with a tetrahydropyrimidine ring: selective α-arylation reaction. Org Biomol Chem 2010;8(10):2465–72.

[154] Fotouhi L, Mosavi M, Heravi MM, Nematollahi D. Efficient electrosynthesis of 1,2,4-triazino[3,4-b]-1,3,4-thiadiazine derivatives. Tetrahedron Lett 2006;47(48):8553–7.

[155] Fakhari AR, Nematollahi D, Shamsipur M, Makarem S, Hosseini Davarani Seyed S, Alizadeh A, et al. Electrochemical synthesis of 5,6-dihydroxy-2-methyl-1-benzofuran-3-carboxylate derivatives. Tetrahedron 2007;63(18):3894–8.

[156] Lin H-H, Chen J-H, Huang C-C, Wang C-J. Apoptotic effect of 3,4-dihydroxybenzoic acid on human gastric carcinoma cells involving JNK/p38 MAPK signaling activation. Int J Cancer 2007;120(11):2306–16.

[157] Nakamura Y, Torikai K, Ohto Y, Murakami A, Tanaka T, Ohigashi H. A simple phenolic antioxidant protocatechuic acid enhances tumor promotion and oxidative stress in female ICR mouse skin: dose-and timing-dependent enhancement and involvement of bioactivation by tyrosinase. Carcinogenesis 2000;21(10):1899–907.

[158] Khalafi L, Rafiee M. Kinetic study of the oxidation and nitration of catechols in the presence of nitrous acid ionization equilibria. J Hazard Mater 2010;174(1–3):801–6.

[159] Nematollahi D, Afkhami A, Mosaed F, Rafiee M. Investigation of the electro-oxidation and oxidation of catechol in the presence of sulfanilic acid. Res Chem Intermed 2004;30 (4–5):299–309.

[160] Gao X-G, Yang C-W, Zhang Z-Z, Zeng C-C, Song X-Q, Hu L-M, et al. Electrochemical oxidation of substituted catechols in the presence of pyrazol-5-ones: characterization of products and reaction mechanism. Tetrahedron 2010;66(52):9880–7.

[161] Moghaddam AB, Kobarfard F, Fakhari AR, Nematollahi D, Davarani SSH. Mechanistic study of electrochemical oxidation of *o*-dihydroxybenzenes in the presence of 4-hydroxy-1-methyl-2(1*H*)-quinolone: application to the electrochemical synthesis. Electrochim Acta 2005;51(4):739–44.

[162] Nematollahi D, Khoshsafar H. Investigation of electrochemically induced Michael addition reactions. Oxidation of some dihydroxybenzene derivatives in the presence of azide ion. Tetrahedron 2009;65(24):4742–50.

[163] Esmaili R, Nematollahi D. Electrochemical oxidation of 4-morpholinoaniline in aqueous solutions: synthesis of a new trimer of 4-morpholinoaniline. Electrochim Acta 2011;56 (11):3899–904.

[164] Fotouhi L, Behrozi L, Heravi MM, Nematollahi D. Electrochemical oxidation of catechols in the presence of pyrimidine-2-thiol: application to electrosynthesis. Phosphorus Sulfur Silicon 2009;184(10):2749–57.

[165] Fakhari AR, Hosseiny Davarani SS, Ahmar H, Makarem S. Electrochemical study of catechols in the presence of 2-thiazoline-2-thiol: application to electrochemical synthesis of new 4,5-dihydro-1,3-thiazol-2-ylsulfanyl-1,2-benzenediol derivatives. J Appl Electrochem 2008;38(12):1743–7.

[166] Nematollahi D, Tammari E. Electrooxidation of 4-methylcatechol in the presence of barbituric acid derivatives. Electrochim Acta 2005;50(18):3648–54.

[167] Chen M, Li X, Ma X. Selective determination of catechol in wastewater at silver doped polyglycine modified film electrode. Int J Electrochem Sci 2012;7:2616–22.

[168] Mersal GA. Electrochemical sensor for voltammetric determination of catechol based on screen printed graphite electrode. Int J Electrochem Sci 2009;4:1167–77.

[169] Kong B, Yin T, Liu X, Wei W. Voltammetric determination of hydroquinone using β-cyclodextrin/poly(*N*-acetylaniline)/carbon nanotube composite modified glassy carbon electrode. Anal Lett 2007;40(11):2141–50.

[170] Kan X, Zhao Q, Zhang Z, Wang Z, Zhu J-J. Molecularly imprinted polymers microsphere prepared by precipitation polymerization for hydroquinone recognition. Talanta 2008;75 (1):22–6.

[171] Kavanoz M, Pekmez N. Poly(vinylferrocenium) perchlorate–polyaniline composite film-coated electrode for amperometric determination of hydroquinone. J Solid State Electrochem 2012;16(3):1175–86.

[172] Zhou K, Zhu Y, Yang X, Li C. Electrocatalytic oxidation of glucose by the glucose oxidase immobilized in graphene–Au–nafion biocomposite. Electroanalysis 2010;22(3):259–64.

[173] Sutherland HS, Higgs KC, Taylor NJ, Rodrigo R. Isobenzofurans and ortho-benzoquinone monoketals in syntheses of xestoquinone and its 9- and 10-methoxy derivatives. Tetrahedron 2001;57(2):309–17.

[174] Nematollahi D, Golabi SM. Investigation of the electromethoxylation reaction part 2: electrochemical study of 3-methylcatechol and 2,3-dihydroxybenzaldehyde in methanol. Electroanalysis 2001;13(12):1008–15.

[175] Nematollahi D, Forooghi Z. Electrochemical oxidation of catechols in the presence of 4-hydroxy-6-methyl-2-pyrone. Tetrahedron 2002;58(24):4949–53.

[176] Nematollahi D, Rahchamani RA. Electro-oxidation of catechols in the presence of benzenesulfinic acid. Application to electro-organic synthesis of new sulfone derivatives. J Electroanal Chem 2002;520(1–2):145–9.

[177] Nematollahi D, Rahchamani R. Electrochemical synthesis of p-tolylsulfonylbenzenediols. Tetrahedron Lett 2002;43(1):147–50.

[178] Nematollahi D, Goodarzi H. Electroorganic synthesis of new benzofuro[2,3-d]pyrimidine derivatives. J Org Chem 2002;67(14):5036–9.

[179] Nematollahi D, Goodarzi H. Electrochemical study of 4-tert-butylcatechol in the presence of 1,3-dimethylbarbituric acid and 1,3-diethyl-2-thiobarbituric acid. Application to the electro-organic synthesis of new corresponding spiropyrimidine derivatives. J Electroanal Chem 2001;517(1–2):121–5.

[180] Nematollahi D, Varmaghani F. Paired electrochemical synthesis of new organosulfone derivatives. Electrochim Acta 2008;53(8):3350–5.

[181] Bard AJ, Faulkner LR. Electrochemical methods. New York: Wiley; 2001. p. 496–500.

[182] Kyriacou DK. Basics of electroorganic synthesis. New York: John Wiley & Sons; 1981. p. 67.

[183] Fotouhi L, Nematollahi D, Heravi MM, Tammari E. An efficient electrochemical method for a unique synthesis of new derivatives of 7*H*-thiazolo[3,2-*b*]-1,2,4-triazin-7-one. Tetrahedron Lett 2006;47(11):1713–16.

[184] Rafiee M, Nematollahi D. Electrochemical oxidation of catechols in the presence of cyanoacetone and methyl cyanoacetate. J Electroanal Chem 2009;626(1–2):36–41.

[185] Nematollahi D, Malakzadeh M. Electrochemical oxidation of quercetin in the presence of benzenesulfinic acids. J Electroanal Chem 2003;547(2):191–5.

[186] Brown DJ. The chemistry of heterocyclic compounds, quinoxalines. In: Taylor EC, Wipf P, editors. Supplement II. Hoboken, New Jersey: Wiley; 2004. p. 251–4.

[187] Sun Z-Y, Botros E, Su A-D, Kim Y, Wang E, Baturay NZ, et al. Sulfoxide-containing aromatic nitrogen mustards as hypoxia-directed bioreductive cytotoxins. J Med Chem 2000;43 (22):4160–8.

[188] Neamati N, Mazumder A, Zhao H, Sunder S, Burke T, Schultz RJ, et al. Diarylsulfones, a novel class of human immunodeficiency virus type 1 integrase inhibitors. Antimicrob Agents Chemother 1997;41(2):385–93.

[189] Hong W-S, Wu C-Y, Lee C-S, Hwang W-S, Chiang MY. Novel iron carbonyl complexes from thiophene-2-carboxaldehyde thiosemicarbazone. J Organomet Chem 2004;689 (2):277–85.

[190] Yamaguchi MU, Barbosa da Silva AP, Ueda-Nakamura T, Dias Filho BP, Conceição da Silva C, Nakamura CV. Effects of a thiosemicarbazide camphene derivative on trichophyton mentagrophytes. Molecules 2009;14(5):1796–807.

[191] Belicchi Ferrari M, Bisceglie F, Pelosi G, Tarasconi P, Albertini R, Pinelli S. New methyl pyruvate thiosemicarbazones and their copper and zinc complexes: synthesis, characterization, X-ray structures and biological activity. J Inorg Biochem 2001;87(3):137–47.

[192] Leite ACL, de Lima RS, Moreira DRdM, Cardoso MVdO, Gouveia de Brito AC, Farias dos Santos LM, et al. Synthesis, docking, and in vitro activity of thiosemicarbazones, aminoacyl-thiosemicarbazides and acyl-thiazolidones against Trypanosoma cruzi. Bioorg Med Chem 2006;14(11):3749–57.

[193] Gielen M. An overview of forty years organotin chemistry developed at the free universities of brussels ULB and VUB. J Braz Chem Soc 2003;14(6):870–7.

[194] Asan A, Isildak I. Determination of major phenolic compounds in water by reversed-phase liquid chromatography after pre-column derivatization with benzoyl chloride. J Chromatogr A 2003;988(1):145–9.

[195] Nagaraja P, Vasantha RA, Sunitha KR. A sensitive and selective spectrophotometric estimation of catechol derivatives in pharmaceutical preparations. Talanta 2001;55 (6):1039–46.

[196] Dong S, Chi L, Yang Z, He P, Wang Q, Fang Y. Simultaneous determination of dihydroxybenzene and phenylenediamine positional isomers using capillary zone electrophoresis coupled with amperometric detection. J Sep Sci 2009;32(18):3232–8.

[197] Li S, Li X, Xu J, Wei X. Flow-injection chemiluminescence determination of polyphenols using luminol–$NaIO_4$–gold nanoparticles system. Talanta 2008;75(1):32–7.

[198] Garcia-Mesa JA, Mateos R. Direct automatic determination of bitterness and total phenolic compounds in virgin olive oil using a pH-based flow-injection analysis system. J Agric Food Chem 2007;55(10):3863–8.

[199] Han L, Zhang X. Simultaneous voltammetry determination of dihydroxybenzene isomers by nanogold modified electrode. Electroanalysis 2009;21(2):124–9.

[200] Guo Q, Huang J, Chen P, Liu Y, Hou H, You T. Simultaneous determination of catechol and hydroquinone using electrospun carbon nanofibers modified electrode. Sens. Actuators, B: Chem 2012;163(1):179–85.

[201] Zhao D-M, Zhang X-H, Feng L-J, Jia L, Wang S-F. Simultaneous determination of hydroquinone and catechol at PASA/MWNTs composite film modified glassy carbon electrode. Colloids Surf, B: Biointerfaces 2009;74(1):317–21.

[202] Ma X, Liu Z, Qiu C, Chen T, Ma H. Simultaneous determination of hydroquinone and catechol based on glassy carbon electrode modified with gold–graphene nanocomposite. Microchim Acta 2013;1–8.

[203] Wang L, Huang P, Bai J, Wang H, Zhang L, Zhao Y. Direct simultaneous electrochemical determination of hydroquinone and catechol at a poly (glutamic acid) modified glassy carbon electrode. Int J Electrochem Soc 2007;2:123–32.

[204] Liu C-C. Electrochemical sensors. Boca Raton, FL: Biomedical Engineering Handbook, CRC Press; 1995. p. 758–63.

[205] Wang Y, Xu H, Zhang J, Li G. Electrochemical sensors for clinic analysis. Sensors 2008;8 (4):2043–81.

[206] Wang J, Li M, Shi Z, Li N, Gu Z. Electrocatalytic oxidation of 3,4-dihydroxyphenylacetic acid at a glassy carbon electrode modified with single-wall carbon nanotubes. Electrochim Acta 2001;47(4):651–7.

[207] Wang J. Carbon-nanotube based electrochemical biosensors: a review. Electroanalysis 2005;17(1):7–14.

[208] Kim SN, Rusling JF, Papadimitrakopoulos F. Carbon nanotubes for electronic and electrochemical detection of biomolecules. Adv Mater 2007;19(20):3214–28.

[209] Wang J, Lin Y. Functionalized carbon nanotubes and nanofibers for biosensing applications. TrAC Trends Anal Chem 2008;27(7):619–26.

[210] Willner I, Yan YM, Willner B, Tel-Vered R. Integrated enzyme-based biofuel cells–a review. Fuel Cells 2009;9(1):7–24.

[211] Gong K, Du F, Xia Z, Durstock M, Dai L. Nitrogen-doped carbon nanotube arrays with high electrocatalytic activity for oxygen reduction. Science 2009;323(5915):760–4.

[212] Shao Y, Liu J, Wang Y, Lin Y. Novel catalyst support materials for PEM fuel cells: current status and future prospects. J Mater Chem 2009;19(1):46–59.

[213] Shao Y, Sui J, Yin G, Gao Y. Nitrogen-doped carbon nanostructures and their composites as catalytic materials for proton exchange membrane fuel cell. Appl Catal, B: Environ 2008;79(1):89–99.

[214] Lin Y, Cui X, Ye X. Electrocatalytic reactivity for oxygen reduction of palladium-modified carbon nanotubes synthesized in supercritical fluid. Electrochem Commun 2005;7 (3):267–74.

[215] Alwarappan S, Erdem A, Liu C, Li C-Z. Probing the electrochemical properties of graphene nanosheets for biosensing applications. J Phys Chem C 2009;113(20):8853–7.

[216] Pumera M, Ambrosi A, Bonanni A, Chng ELK, Poh HL. Graphene for electrochemical sensing and biosensing. TrAC Trends Anal Chem 2010;29(9):954–65.

[217] Fan Y, Lu H-T, Liu J-H, Yang C-P, Jing Q-S, Zhang Y-X, et al. Hydrothermal preparation and electrochemical sensing properties of TiO_2–graphene nanocomposite. Colloids Surf, B: Biointerfaces 2011;83(1):78–82.

[218] Li L, Du Z, Liu S, Hao Q, Wang Y, Li Q, et al. A novel nonenzymatic hydrogen peroxide sensor based on MnO_2/graphene oxide nanocomposite. Talanta 2010;82(5):1637–41.

[219] Jin E, Lu X, Cui L, Chao D, Wang C. Fabrication of graphene/prussian blue composite nanosheets and their electrocatalytic reduction of H_2O_2. Electrochim Acta 2010;55 (24):7230–4.

[220] Yin H, Zhou Y, Ma Q, Ai S, Chen Q, Zhu L. Electrocatalytic oxidation behavior of guanosine at graphene, chitosan and Fe_3O_4 nanoparticles modified glassy carbon electrode and its determination. Talanta 2010;82(4):1193–9.

[221] Salimi A, Hallaj R. Adsorption and reactivity of chlorogenic acid at a hydrophobic carbon ceramic composite electrode: application for the amperometric detection of hydrazine. Electroanalysis 2004;16(23):1964–71.

[222] Lawrence NS, Davis J, Compton RG. Electrochemical detection of thiols in biological media. Talanta 2001;53(5):1089–94.

[223] White PC, Lawrence NS, Tsai YC, Davis J, Compton RG. Electrochemically driven derivatisation-detection of cysteine. Microchim Acta 2001;137(1–2):87–91.

[224] Hugo Seymour E, Lawrence NS, Beckett EL, Davis J, Compton RG. Electrochemical detection of aniline: an electrochemically initiated reaction pathway. Talanta 2002;57 (2):233–42.

[225] Xu Z, Chen X, Qu X, Dong S. Electrocatalytic oxidation of catechol at multi-walled carbon nanotubes modified electrode. Electroanalysis 2004;16(8):684–7.

[226] Lin C-H, Sheu J-Y, Wu H-L, Huang Y-L. Determination of hydroquinone in cosmetic emulsion using microdialysis sampling coupled with high-performance liquid chromatography. J Pharm Biomed Anal 2005;38(3):414–19.

[227] Chen GN, Liu JS, Duan JP, Chen HQ. Coulometric detector based on porous carbon felt working electrode for flow injection analysis. Talanta 2000;53(3):651–60.

[228] Pistonesi MF, Di Nezio MS, Centurion ME, Palomeque ME, Lista AG, Fernández Band BS. Determination of phenol, resorcinol and hydroquinone in air samples by synchronous fluorescence using partial least-squares (PLS). Talanta 2006;69(5):1265–8.

[229] Li S-J, Xing Y, Wang G-F. A graphene-based electrochemical sensor for sensitive and selective determination of hydroquinone. Microchim Acta 2012;176(1–2):163–8.

[230] Yiyi S, Yougen T, Hongtao L, Ping H. Electrochemical determination of hydroquinone using hydrophobic ionic liquid-type carbon paste electrodes. Chem Cent J 2010;4:17–24.

[231] Timur S, Pazarlioğlu N, Pilloton R, Telefoncu A. Detection of phenolic compounds by thick film sensors based on Pseudomonas putida. Talanta 2003;61(2):87–93.

[232] Ghanem MA. Electrocatalytic activity and simultaneous determination of catechol and hydroquinone at mesoporous platinum electrode. Electrochem Commun 2007;9 (10):2501–6.

[233] Ahammad AJS, Rahman MM, Xu G-R, Kim S, Lee J-J. Highly sensitive and simultaneous determination of hydroquinone and catechol at poly(thionine) modified glassy carbon electrode. Electrochim Acta 2011;56(14):5266–71.

[234] Zhang H, Zhao J, Liu H, Liu R, Wang H, Liu J. Electrochemical determination of diphenols and their mixtures at the multiwall carbon nanotubes/poly(3-methylthiophene) modified glassy carbon electrode. Microchim Acta 2010;169(3–4):277–82.

[235] Bu C, Liu X, Zhang Y, Li L, Zhou X, Lu X. A sensor based on the carbon nanotubes-ionic liquid composite for simultaneous determination of hydroquinone and catechol. Colloids Surf, B: Biointerfaces 2011;88(1):292–6.

[236] Wang L, Huang P, Bai J, Wang H, Zhang L, Zhao Y. Simultaneous electrochemical determination of phenol isomers in binary mixtures at a poly(phenylalanine) modified glassy carbon electrode. Int J Electrochem Soc 2006;1:403–13.

[237] Wang Z, Li S, Lv Q. Simultaneous determination of dihydroxybenzene isomers at single-wall carbon nanotube electrode. Sens. Actuators, B: Chem 2007;127(2):420–5.

[238] Kong Y, Chen X, Yao C, Ma M, Chen Z. A voltammetric sensor based on electrochemically activated glassy carbon electrode for simultaneous determination of hydroquinone and catechol. Anal Methods 2011;3(9):2121–6.

[239] Bai J, Guo L, Ndamanisha J, Qi B. Electrochemical properties and simultaneous determination of dihydroxybenzene isomers at ordered mesoporous carbon-modified electrode. J Appl Electrochem 2009;39(12):2497–503.

[240] Sun D, Zhang H. Electrochemical determination of acetaminophen using a glassy carbon electrode coated with a single-wall carbon nanotube-dicetyl phosphate film. Microchim Acta 2007;158(1–2):131–6.

[241] Yang X-X, Hu Z-P, Duan W, Zhu Y-Z, Zhou S-F. Drug–herb interactions: eliminating toxicity with hard drug design. Curr Pharm Des 2006;12(35):4649–64.

[242] Wan Q, Wang X, Yu F, Wang X, Yang N. Poly(taurine)/MWNT-modified glassy carbon electrodes for the detection of acetaminophen. J Appl Electrochem 2009;39(6):785–90.

[243] <http://www.assistpainrelief.com/dyn/304/Paracetamol.html>, accessed on 1.9.2012.

[244] Lourenção BC, Medeiros RA, Rocha-Filho RC, Mazo LH, Fatibello-Filho O. Simultaneous voltammetric determination of paracetamol and caffeine in pharmaceutical formulations using a boron-doped diamond electrode. Talanta 2009;78(3):748–52.

[245] Özcan A, Şahin Y. A novel approach for the determination of paracetamol based on the reduction of *N*-acetyl-*p*-benzoquinoneimine formed on the electrochemically treated pencil graphite electrode. Anal Chim Acta 2011;685(1):9–14.

[246] Sirajuddin, Khaskheli AR, Shah A, Bhanger MI, Niaz A, Mahesar S. Simpler spectrophotometric assay of paracetamol in tablets and urine samples. Spectrochim Acta, A: Mol Biomol Spectrosc 2007;68(3):747–51.

[247] Baptistao M, Rocha WFdC, Poppi RJ. Quality control of the paracetamol drug by chemometrics and imaging spectroscopy in the near infrared region. J Mol Struct 2011;1002 (1–3):167–71.

[248] Gioia MG, Andreatta P, Boschetti S, Gatti R. Development and validation of a liquid chromatographic method for the determination of ascorbic acid, dehydroascorbic acid and acetaminophen in pharmaceuticals. J Pharm Biomed Anal 2008;48(2):331–9.

[249] Selvan P, Gopinath R, Saravanan V, Gopal N, Kumar A, Periyasamy K. Simultaneous estimation of paracetamol and aceclofenac in combined dosage forms by RP-HPLC method. Asian J Chem 2007;19:1004–10.

[250] Ishii Y, Iijima M, Umemura T, Nishikawa A, Iwasaki Y, Ito R, et al. Determination of nitrotyrosine and tyrosine by high-performance liquid chromatography with tandem mass

spectrometry and immunohistochemical analysis in livers of mice administered acetaminophen. J Pharm Biomed Anal 2006;41(4):1325–31.

[251] Umasankar Y, Unnikrishnan B, Chen S-M, Ting T-W. Effective determination of acetaminophen present in pharmaceutical drug using functionalized multi-walled carbon nanotube film. Int J Electrochem Soc 2012;7:484–98.

[252] Habibi B, Jahanbakhshi M, Pournaghi-Azar MH. Differential pulse voltammetric simultaneous determination of acetaminophen and ascorbic acid using single-walled carbon nanotube-modified carbon–ceramic electrode. Anal Biochem 2011;411(2):167–75.

[253] Su W-Y, Cheng S-H. Electrochemical oxidation and sensitive determination of acetaminophen in pharmaceuticals at poly(3,4-ethylenedioxythiophene)-modified screen-printed electrodes. Electroanalysis 2010;22(6):707–14.

[254] Ding Y, Garcia CD. Determination of nonsteroidal anti-inflammatory drugs in serum by microchip capillary electrophoresis with electrochemical detection. Electroanalysis 2006;18 (22):2202–9.

[255] Lawrence D. Retrospective data strengthen Alzheimer's link with aspirin and NSAIDs. Lancet 2002;360(9338):1003.

[256] Weggen S, Rogers M, Eriksen J. NSAIDs: small molecules for prevention of Alzheimer's disease or precursors for future drug development? Trends Pharmacol Sci 2007;28 (10):536–43.

[257] Cuzick J, Otto F, Baron JA, Brown PH, Burn J, Greenwald P, et al. Aspirin and nonsteroidal anti-inflammatory drugs for cancer prevention: an international consensus statement. Lancet Oncol 2009;10(5):501–7.

[258] Maree AO, Curtin RJ, Dooley M, Conroy RM, Crean P, Cox D, et al. Platelet response to low-dose enteric-coated aspirin in patients with stable cardiovascular disease. J Am Coll Cardiol 2005;46(7):1258–63.

[259] Poulsen TS, Kristensen SR, Korsholm L, Haghfelt T, Jørgensen B, Licht PB, et al. Variation and importance of aspirin resistance in patients with known cardiovascular disease. Thromb Res 2007;120(4):477–84.

[260] Wudarska E, Chrzescijanska E, Kusmierek E, Rynkowski J. Voltammetric studies of acetylsalicylic acid electrooxidation at platinum electrode. Electrochim Acta 2013;93 (0):189–94.

[261] Zen J-M, Ting Y-S, Shih Y. Voltammetric determination of caffeine in beverages using a chemically modified electrode. Analyst 1998;123(5):1145–7.

[262] Spătaru N, Sarada BV, Tryk DA, Fujishima A. Anodic voltammetry of xanthine, theophylline, theobromine and caffeine at conductive diamond electrodes and its analytical application. Electroanalysis 2002;14(11):721–8.

[263] Yamauchi Y, Nakamura A, Kohno I, Kitai M, Hatanaka K, Tanimoto T. Simple and rapid UV spectrophotometry of caffeine in tea coupled with sample pre-treatment using a cartridge column filled with polyvinylpolypyrrolidone (PVPP). Chem Pharm Bull 2008;56 (2):185–8.

[264] Emre D, Özaltın N. Simultaneous determination of paracetamol, caffeine and propyphenazone in ternary mixtures by micellar electrokinetic capillary chromatography. J Chromatogr B: Anal Technol Biomed Life Sci 2007;847(2):126–32.

[265] Mersal GM. Experimental and computational studies on the electrochemical oxidation of caffeine at pseudo carbon paste electrode and its voltammetric determination in different real samples. Food Anal Methods 2012;5(3):520–9.

[266] Faria EO, Lopes Junior ACV, Souto DEP, Leite FRF, Damos FS, de Cássia Silva Luz R, et al. Simultaneous determination of caffeine and acetylsalicylic acid in pharmaceutical

formulations using a boron-doped diamond film electrode by differential pulse voltammetry. Electroanalysis 2012;24(5):1141–6.

[267] Sanghavi BJ, Srivastava AK. Simultaneous voltammetric determination of acetaminophen, aspirin and caffeine using an in situ surfactant-modified multiwalled carbon nanotube paste electrode. Electrochim Acta 2010;55(28):8638–48.

[268] Vadukumpully S, Paul J, Mahanta N, Valiyaveettil S. Flexible conductive graphene/poly (vinyl chloride) composite thin films with high mechanical strength and thermal stability. Carbon NY 2011;49(1):198–205.

[269] Izutsu K. Potentiometry in non-aqueous solutions. Electrochemistry in nonaqueous solutions. Weinheim: Wiley-VCH Verlag GmbH & Co. KGaA; 2003. p. 167–200.

[270] Fakhari AR, Nematollahi D, Moghaddam AB. Mechanistic study of electrochemical oxidation of catechols in the presence of 4-hydroxy-1-methyl-2(1*H*)-quinolone: application to the electrochemical synthesis. Electrochim Acta 2005;50(27):5322–8.

[271] Dabaghi HH, Moghaddam AB, Kazemzad M, Dinarvand R, Aryanasab F, Nabid MR. A strategy for the electro-organic synthesis of new hydrocaffeic acid derivatives. J Appl Electrochem 2008;38(3):409–13.

[272] Nematollahi D, Shayani-Jam H, Alimoradi M, Niroomand S. Electrochemical oxidation of acetaminophen in aqueous solutions: kinetic evaluation of hydrolysis, hydroxylation and dimerization processes. Electrochim Acta 2009;54(28):7407–15.

[273] Raoof JB, Ojani R, Amiri-Aref M, Chekin F. Catechol as an electrochemical indicator for voltammetric determination of *N*-acetyl-l-cysteine in aqueous media at the surface of carbon paste electrode. J Appl Electrochem 2010;40(7):1357–63.

[274] Nematollahi D, Rafiee M. Electrochemical oxidation of catechols in the presence of acetylacetone. J Electroanal Chem 2004;566(1):31–7.

[275] Crews, P., Organic structure analysis. 1998, New York, Oxford: p. 63,331,332, 333,336.

[276] Bellami LJ. The infra-red spectra of complex molecules. 3rd Edition London: Chapman and Hall; 1975.

[277] Bruice PY. Organic chemistry. 2nd Edition United States: Pearson Education Inc.; 2004.

[278] Abo Aly MM. Infrared and Raman spectra of some symmetric azines. Spectrochim Acta, A: Mol Biomol Spectrosc 1999;55(9):1711–14.

[279] Affan MA, Wan FS, Ngaini Z, Shamsuddin M. Synthesis, characterization and biological studies of organotin (iv) complexes of thiosemicarbazone ligand derived from pyruvic acid: x-ray crystal structure of [ME_2 SN (PAT)]. Malays J Anal Sci 2009;13(1):63–72.

[280] <http://www.chem.utoronto.ca/coursenotes/analsci/stats/ftest.html>, accessed on 10.2.2012.

[281] <http://www.chem.utoronto.ca/coursenotes/analsci/stats/ttest.html>, accessed on 10.2.2012.

[282] White PC, Lawrence NS, Davis J, Compton RG. Electrochemically initiated 1, 4 additions: a versatile route to the determination of thiols. Anal Chim Acta 2001;447(1):1–10.

[283] Luo J, Jiang S, Zhang H, Jiang J, Liu X. A novel non-enzymatic glucose sensor based on Cu nanoparticle modified graphene sheets electrode. Anal Chim Acta 2012;709(0):47–53.

[284] Fan H, Li Y, Wu D, Ma H, Mao K, Fan D, et al. Electrochemical bisphenol A sensor based on N-doped graphene sheets. Anal Chim Acta 2012;711(0):24–8.

[285] Tehrani RMA, Ab Ghani S. MWCNT-ruthenium oxide composite paste electrode as non-enzymatic glucose sensor. Biosens Bioelectron 2012;38(1):278–83.

[286] Reddy ALM, Ramaprabhu S. Nanocrystalline metal oxides dispersed multiwalled carbon nanotubes as supercapacitor electrodes. J Phys Chem C 2007;111(21):7727–34.

[287] Estaline Amitha F, Leela Mohana Reddy A, Ramaprabhu S. A non-aqueous electrolyte-based asymmetric supercapacitor with polymer and metal oxide/multiwalled carbon nanotube electrodes. J Nanopart Res 2009;11(3):725–9.

[288] Song H-K, Hwang H-Y, Lee K-H, Dao LH. The effect of pore size distribution on the frequency dispersion of porous electrodes. Electrochim Acta 2000;45(14):2241–57.

[289] Zhang K, Zhang LL, Zhao XS, Wu J. Graphene/polyaniline nanofiber composites as supercapacitor electrodes. Chem Mater 2010;22(4):1392–401.

[290] Ghadimi H, Ali ASM, Mohamed N, Ab Ghani S. Electrochemical oxidation of catechol in the presence of thiosemicarbazide. J Electrochem Soc 2012;159(6):E127–31.

[291] Du H, Ye J, Zhang J, Huang X, Yu C. A voltammetric sensor based on graphene-modified electrode for simultaneous determination of catechol and hydroquinone. J Electroanal Chem 2011;650(2):209–13.

[292] Ghorbani-Bidkorbeh F, Shahrokhian S, Mohammadi A, Dinarvand R. Simultaneous voltammetric determination of tramadol and acetaminophen using carbon nanoparticles modified glassy carbon electrode. Electrochim Acta 2010;55(8):2752–9.

[293] Kumar SA, Tang C-F, Chen S-M. Electroanalytical determination of acetaminophen using nano-TiO_2/polymer coated electrode in the presence of dopamine. Talanta 2008;76 (5):997–1005.

[294] Zidan M, Tan W, Abdullah AH, Zainal Z, Goh JK. Electrocatalytic oxidation of paracetamol mediated by lithium doped microparticles Bi_2O_3/MWCNT modified electrode. Asian J Chem 2011;23(7):3029–32.

[295] Li C, Zhan G, Yang Q, Lu J. Electrochemical investigation of acetaminophen with a carbon nano-tube composite film electrode. Bulletin of the Korean Chemical Society 2006;27 (11):1854–60.

[296] Sotomayor MD, Sigoli A, Lanza MR, Tanaka AA, Kubota LT. Construction and application of an electrochemical sensor for paracetamol determination based on iron tetrapyridinoporphyrazine as a biomimetic catalyst of P450 enzyme. J Braz Chem Soc 2008;19 (4):734–43.

[297] Dalmasso PR, Pedano ML, Rivas GA. Electrochemical determination of ascorbic acid and paracetamol in pharmaceutical formulations using a glassy carbon electrode modified with multi-wall carbon nanotubes dispersed in polyhistidine. Sens. Actuators, B: Chem 2012;173 (0):732–6.

[298] Lu T-L, Tsai Y-C. Sensitive electrochemical determination of acetaminophen in pharmaceutical formulations at multiwalled carbon nanotube-alumina-coated silica nanocomposite modified electrode. Sens. Actuators, B: Chem 2011;153(2):439–44.

[299] ShangGuan X, Zhang H, Zheng J. Electrochemical behavior and differential pulse voltammetric determination of paracetamol at a carbon ionic liquid electrode. Chin J Anal Chem 2008;391(3):1049–55.

[300] Atta NF, Galal A, Azab SM. Electrochemical determination of paracetamol using gold nanoparticles—application in tablets and human fluids. Int J Electrochem Soc 2011;6:5082–96.

[301] Kachoosangi RT, Wildgoose GG, Compton RG. Sensitive adsorptive stripping voltammetric determination of paracetamol at multiwalled carbon nanotube modified basal plane pyrolytic graphite electrode. Anal Chim Acta 2008;618(1):54–60.

[302] Xu F, Ru H-Y, Sun L-X, Zou Y-J, Jiao C-L, Wang T-Y, et al. A novel sensor based on electrochemical polymerization of diglycolic acid for determination of acetaminophen. Biosens Bioelectron 2012;38(1):27–30.

[303] Li L-J, Zhong L, Cai Z, Cheng H, Yu L-B. Electrochemical behaviors and determination of paracetamol at multi-walled carbon nanotubes and L-cysteine co-assembling modified gold electrode. Chin J Anal Chem 2008;36(12):1651–6.

[304] Hou X, Shen G, Meng L, Zhu L, Guo M. Multi-walled carbon nanotubes modified glass carbon electrode and its electrocatalytic activity towards oxidation of paracetamol. Russ J Electrochem 2011;47(11):1262–7.

[305] Noviandri I, Rakhmana R. Carbon paste electrode modified with carbon nanotubes and poly(3-aminophenol) for voltammetric determination of paracetamol. Int J Electrochem Soc 2012;7:4479–87.

[306] Wang S-F, Xie F, Hu R-F. Carbon-coated nickel magnetic nanoparticles modified electrodes as a sensor for determination of acetaminophen. Sens. Actuators, B: Chem 2007;123 (1):495–500.

[307] Li M, Jing L. Electrochemical behavior of acetaminophen and its detection on the PANI–MWCNTs composite modified electrode. Electrochim Acta 2007;52(9):3250–7.

[308] Zidan M, Tee TW, Abdullah AH, Zainal Z, Kheng GJ. Electrochemical oxidation of paracetamol mediated by nanoparticles bismuth oxide modified glassy carbon electrode. Int J Electrochem Soc 2011;6:279–88.

[309] Sánchez-Obrero G, Mayén M, Mellado JMR, Rodríguez-Amaro R. Electrocatalytic oxidation of acetaminophen on a PVC/TTFTCNQ composite electrode modified by gold nanoparticles: application as an amperometric sensor. Int J Electrochem Soc 2001;6:2001–11.

[310] Ghadimi H, Nasiri-Tabrizi B, Moozarm Nia P, Jefrey Basirun W, Tehrani RMA, Lorestani F. Nanocomposites of nitrogen-doped graphene decorated with a palladium silver bimetallic alloy for use as a biosensor for methotrexate detection. RSC Adv 2015;5:99555–65.

Printed in the United States
By Bookmasters